EVOLUTIONARY PROGRESS

EVOLUTION AS PROGRESS

EVOLUTIONARY PROGRESS

Edited by
Matthew H. Nitecki

The University of Chicago Press
Chicago and London

Evolutionary progress

MATTHEW H. NITECKI is curator of fossil invertebrates in the
Department of Geology at the Field Museum of Natural
History and a member of the Committee on Evolutionary
Biology at the University of Chicago.

The University of Chicago Press, Chicago 60637
The University of Chicago Press, Ltd., London

© 1988 by The University of Chicago
All rights reserved. Published 1988
Printed in the United States of America
97 96 95 94 93 92 91 90 89 88   54321

Library of Congress Cataloging in Publication Data

Evolutionary progress.

  Includes bibliographical references and index.
  1. Evolution. I. Nitecki, Matthew H.
QH371.E929   1989   575   88-20835
ISBN 0-226-58692-8
ISBN 0-226-58693-6 (pbk.)

Table of Contents

Preface

Pessimism prevails in our world, and the whole fabric of our society challenges our expectations. Jacques Monod explains the purpose of life by chance, Paul Kennedy prophesies that the rise of the great powers inevitably leads to their downfall, while Gunther Stent infers that God is dead, art is dead, music has reached its zenith and science has reached its limit. Despair seems the fashion of the day: it presumes that population growth is a deadly bomb; that shortages of food will produce mass hunger; that AIDS will decimate our population; that mass poverty will rule tomorrow; that religious and fundamentalist upheavals will destroy freedom; that the present monetary system will collapse and rupture social order; that commercialism is omnipresent and ideas dead; that pollution will smother us and lack of fuels will freeze us; and that Tartars will light the torches of war with megatons.

However, merely to survive in our daily life we must have faith in tomorrow. Without such faith in the future, we cannot succeed in our technological, social, and educational enterprises. We need to keep it, although in many empirical senses we cannot justify doing so. Its absence spells despair. Thus the idea of progress cannot be dead, but what does exist is our changing understanding of its meaning.

The concept of progress as applied to the accumulation of knowledge, for example, about the development of transportation or a pair of skis, can perhaps be measured but is a very different thing from the demonstration of progressive change in evolution. The problem seems to be in the application of the same term to a different concept, or conversely, giving different meanings to the same word.

In the nineteenth-century biological literature, including Darwin's *Origin of Species*, the words *progress* and *evolution* were often tied together, and *laws of progress* were generally believed to be characteristic of both biological and social evolution. More recently, with the increasing importance of microevolutionary theory and extrapolations from it, many have felt it

vii

necessary to distinguish the consequences of neo-Darwinian natural selection from the older *progressivist* theories of social evolution, and to clarify misconceptions surrounding notions of biological progress. As a result, the concept of progress has been all but banned from evolutionary biology as being anthropocentric or at best of limited and ambiguous usefulness. Within the last few decades, however, there has been an explosion of interest in adaptational complexity, long-range evolutionary trends, and developing theories of evolution above the species level, and with them, in concepts like *hierarchy* and *levels of organization*, which are at least potentially suggestive of *progressivist* perspectives.

This volume examines the history and conceptual foundations of the biological idea of progress, and reexamines whether or not the exclusion of notions of progress from evolutionary biology is justifiable in light of recent theoretical developments. It brings together leading philosophers, historians, and scientists from various disciplines to promote a comprehensive understanding, if not a consensus, on these issues. We consider what is known about biochemical evolution and genetic regulation, developmental constraints, the evolution of levels of organization, the existence of different levels and units of selection, nonequilibrium thermodynamics, macroevolution, and trends in morphology or speciation through time. We ask again whether or not some form of progress or improvement is manifested, or whether evolution might once have been progressive but has now reached a steady state of progressive adaptation countered by mutational or environmental degradation, or if it has always been nothing more than a simple random walk.

This book is the result of the Spring Systematics Symposium held in May, 1987, at the Field Museum in Chicago. I am grateful to Garland Allen, William Burger, Richard Burkhardt, Brian Charlesworth, Joel Cracraft, Karl Flessa, Jack Fooden, David Hull, David Joravsky, Richard Klein, Scott Lidgard, David Raup, Robert Richards, Philip Sloan, George Stocking, John Terrell, James Valentine, Michael Wade, William Wimsatt, and Bertram Woodland for their cooperation, reviews of individual chapters, and for their evolutionary advice. Zbigniew Jastrzebski prepared many text-figures. Doris Nitecki was responsible for the grammar, appearance, and camera-ready preparation of the book; she side-by-side nursed the entire project from its conception to its final stage. I thank you all, for your learned counsel.

INTRODUCTION

Discerning the Criteria for Concepts of Progress

Matthew H. Nitecki

It was Sir Arthur Eddington who coined the term "time's arrow." To him the universe is running down. "Running from past to future is the doing of an event; the same sequence running from future to past is the undoing of it" (1935, p. 74). He claims to be an evolutionist and hence he follows the second law of thermodynamics. But is he right, this magnificent writer of almost poetic prose of physics and astronomy? Can the second law be unquestionably applicable to the universe at large, or to living processes? Are not astronomers the most outrageous science fiction writers, applying the findings of physicists to the whole known and unknown universe? Perhaps they are right, but so universal and uncritical an application of the second law may be, perhaps, not wholly justifiable. We must remember, however, that laws in physics, and in biology, only describe what occurs statistically. "Natural laws" do not compel events to occur.

Biology, particularly evolutionary biology, is a historical science, and as such is assumed to have an arrow of time. But which way is the arrow pointing, if there is an arrow at all? Is life running down, as some proponents of the second law believe, or is the evidence of progress in the history of life real? Can, for example, the "innovation" of human housing be traced from animal nests through caves to shelter structures of prehuman societies? And if so, are our spacious, clean and pollen-free houses a document of progress? And as a consequence will tomorrow's houses be even "better"? Is our faith in the future justified? These intriguing, philosophical and endless questions have been asked by philosophers, written about by historians, argued over by anthropologists, and speculated on by biologists.

The most important consideration of the present volume is *whether biological organisms have evolved progressively, and if so, how the process*

3

may be characterized. Often contemporary biologists will deny evolutionary progress among biological organisms but will admit our own progressive accumulation of scientific and technological knowledge. But isn't there a conflict here? For if man is part of nature, are we allowed to consider human social progress in isolation and apart from animal progress? If we agree that progress is real only for *Homo sapiens*, then we assume the uniqueness of man and his separation from the animal kingdom. If, on the other hand, we think of ourselves as part of the animal kingdom, do we then agree that progress exists for *all* life or that it does not exist at all?

Philosophy of the Idea

How nonbiologists "philosophize" about evolutionary progress is best exemplified by two professional philosophers, Bertrand Russell and Henri Bergson. Russell wrote much on progress (e.g., 1927, 1935, 1944, 1945) and is the greatest opponent of evolutionary progress. Any thoughtful social change, Russell argues, must be based on scientific deliberations, but whether such, or any other, change will be for the better or for the worse does not depend on science (1944, p. 726). An airplane ticket is useful whether you fly to perform heart surgery or smuggle heroin.

Russell criticized the doctrine of Cosmic Purpose, which claims that evolution has a direction "towards something ethically valuable" (1935, p. 190). Cosmic purpose has theistic, pantheistic and emergent forms. In the theistic form, purpose exists in the mind of God; in the pantheistic, progress is in the universe's creative force; and in the emergent, progress is in the blind impulsion that leads to ever higher and higher forms. Russell agrees that species probably evolved through natural selection from preexisting species. But these past events tell him *nothing about the future.* The geologic history of the earth or the paleontological interpretation of the fossil record does not anywhere support the idea of progress. But even if it would, it would not prove that progress is the property of the universe. From the history of life on our planet, we cannot possibly assume the existence of progress in the galaxy. What is only certain is change and continuity. But neither change nor continuity is progress, they are the properties of the whole universe and not only of the crust of our planet. Therefore, cosmic purpose is not the subject matter of biology or philosophy -- but a mere wish fulfillment.

The progressive form of evolutionism is best presented by Bergson (Russell 1935, pp. 191-92). To biologists, Bergson is known for his poetic and beautifully written book (1907) on creative evolution and vital force. While his general philosophy may be respected in some circles, his biology and his ideas of progress are totally erroneous. (For a review of Bergson's philosophy, see Russell 1945, pp. 791-810.) Nevertheless, certain of his ideas were clear and appear timely now, as, for example: "Darwin spoke of very slight variations being accumulated by natural selection. He was not ignorant of the facts of sudden variations . . . . Such is still the opinion of many naturalists. It is tending, however, to give way to the opposite [and, of course, still problematic and not yet defended] idea that a new species comes into being all at once by the simultaneous appearance of several new characters . . . [a] hypothesis, already proposed by various authors . . . . Species pass through alternate periods of stability and transformation . . ." (1907, p. 71).

To Bergson, evolution and "all nature is the expression of the creative urgency of the fulfillment of novel ends created in the process of time itself" (p. xiv). Organisms must be defined not by any special morphologies or characters but by the tendencies to emphasize them (p. 118). "Each species, each individual even, retains only a certain impetus from the universal vital impulsion and tends to use this energy in its own interest. In this consists *adaptation*" (pp. 57-58). All the genuine novelties in evolution could not have been predicted, but since the past never dies, life becomes richer by retaining this past.

Bergson, unlike other biologists, was not interested in finding out the succession of life, or of individual fossil taxa, but in deciphering the principal direction of evolution in general. But of all the directions only that leading to man was important to him. After all, infusorias and amoebas are *rudimentary organisms* with only traces of effective psychological activities (p. 1). The progress of organic development is controlled by latent or potential will in a stream of life which, like a wind, divides itself at the street corner. This potential will is a vital action, or *élan vital*. "Living being is above all a thoroughfare, and . . . the essence of life is in the movement by which life is transmitted" (p. 142). The process from amoeba to man means an increase in mobility, leading to consciousness, and to a future of higher mobility and instinct, but not higher intelligence [sic]. Bergson's evolutionary process obviously has a vector -- it looks to man: "Man might be considered the reason for the existence of the entire organization of life on our planet" (p. 203).

Of the authors in the philosophy section of this book, David L. Hull and William B. Provine are as critical of the idea of progress as Russell is. Michael Ruse claims that faith in progress has deep roots among evolutionary biologists, and Francisco J. Ayala makes cautious use of the limited concept of progress. Hull comes superbly qualified to analyze the problem of progress and is one of its most exacting critics. To him science, like all other human activities, is subject to the cultural fashions of the day. He takes exception to the conventional biological wisdom that the fossil record through geological time exhibits directionality in the general increase of biomass and its complexity. He is also skeptical that an increase in biomass must accompany a general directional increase in diversity. Not only does Hull not believe in progress, but he sees direction as theory-laden, just as progress is theory-laden. None of the published plots of progress can be definitely shown to have any direction. Many species claimed in the past to have been "good" species may not be species at all. The observed "progressive" trends may be derived largely from the sets of classifications used. The evolutionary convention of organizing data from primitive to advanced forces us to construct our classification from protozoa to higher vertebrates -- which in turn is superimposed on our phylogeny.

Hull questions the cherished concept of progress; nothing after all can grow exponentially forever, and the expansion in one area can only occur at the expense of another. Sooner or later, a steady state must be reached in the long history of life only to be punctuated by mass extinctions. Hull also challenges the new concept of directionality, which replaces progress and inquires how there could be any progress without directional change.

Furthermore, Hull argues that progress has been postulated only by assuming hypothetical "higher levels." To the believers in progress this is the emergence of man, and the higher levels, of course, are us. Since we are rational, there is progress because rationality has evolved; we have culture and civilization and, therefore, because culture and civilization have evolved there is progress. Hull's examination of the 4.5 billion years of the history of earth does not reveal any evidence of a direction or a tendency toward either rationality or culture, except for the most recent archaeological record of *Homo sapiens*.

Provine, a historian of modern biology, is very deeply interested in the mechanics of evolution and claims for design in nature. Like Hull he is critical of quick and superficial conclusions, and through his historical analyses he denies the existence of a purposive force in nature responsible for the evolutionary path of life. Provine argues for a purely mechanistic interpretation of the evolution of life, as postulated by Charles Darwin and others.

Provine's cosmic view, reminiscent of Russell's (1927) forceful free man's worship, is that the universe is indifferent to life. Either the universe is expanding and life will end in a heat-death, or the universe is pulsating and life will crush into insignificance. In the long perspective of geological history the short duration of human life is equally meaningless, therefore there is no *ultimate* meaning to the human life. Provine distinguishes between the real progress in knowledge of life and the assumed progress of evolution itself. He shows how the mechanistic interpretation of life wins the day and makes assumptions of design, purpose or progress in history of life preposterous. There are no gods, no designers and no progress in nature.

Ayala has contributed to our understanding of the definition of and possible meaning for progress. At first progress appears as a clear-cut idea, but when pondered on, it appears more complex. Progress generally implies an improvement, a growth or movement forward toward a goal. But improvement means that things were not so good before. Who on earth can say they will improve tomorrow?

Ayala wishes to clear away all the semantic overburden from the concept of biological progress and to remove from it association with evolution, change, direction and trend. Progress can be either *descriptive* or *axiological*; it must sustain the occurred change, and this change is an improvement or betterment; progress is thus a "directional change toward the better." To Ayala continuity of change produces either a *uniform* progress in which a later member is better than a preceding, or a *net* progress in which a later member is, *on the average*, better than a preceding. On the other hand, the scope of the sequence produces a *general* progress when all the historical sequences and their durations are better, or a *particular* progress when some sequences at certain times are better.

Ayala's examination of the history of life shows no uniform progress; however, he sees the general net progress in the expansion of life, particularly in the expansion of systematic diversity, in the increase of the number of individuals, and in the increase of the biomass and energy flow. To Ayala, particular net progress can be shown in the evolution of life. Moreover, in their abilities to obtain and process information, humans are the most progressive organisms.

Ruse shows that there is in evolutionary biology a "bad" trend and a "good" trend. He has explained these terms before (Ruse 1986), and here he develops them further. The bad trend, which attempts to describe evolution as the most progressive doctrine, is associated with the views of Herbert Spencer. It sees evolution as a progressive phenomenon, from "least" to "best," with the peak occupied by us humans. The good trend, associated with the views of Darwin, sees evolution as a nonprogressive, directionless process with no "better" or "worse." Ruse does not argue for progress in evolution but rather that Darwin himself was a progressionist. Ruse agrees with Richards (below), but he goes further than that; his examination of the history of evolutionary biology, particularly the contemporary ideas, demonstrates to him that many of today's well-known evolutionists are progressionists also. He examines the idea of progress among the founders of the synthetic theory -- Julian Huxley, Ronald Fisher, George Gaylord Simpson, G. Ledyard Stebbins and Theodosius Dobzhansky -- and finds them all to be progressionists. But Ruse surprises us most by his in-depth analysis of the writings of such "reluctant progressionists" as Edward O. Wilson, Richard Dawkins and Stephen Jay Gould.

Ideas of progress, as advocated by these present-day evolutionists, are neither easily read nor openly Spencerian, but nevertheless they are, according to Ruse, persistent and value-laden notions. There are three reasons for this persistence. First, we as humans are not disinterested observers; we are products, and even for our understanding, by-products of evolution. This is an idea with which most contributors to the present volume agree. Second, the mechanisms of adaptations are almost "design-like" as if created by intentions. Third, evolutionary theory is science and, as in all sciences, we the scientists see unmistakable progress in evolutionary biology also. Thus, we read progress into everything, including the history of life itself.

Historical and Comparative Approaches

From the large historical literature, the most recent history of progress is by the sociologist-historian Robert Nisbet (1980). In his broadly humanistic book, the origin of the idea is traced to classical antiquity rather than to the sixteenth century as claimed by Bury (1932). Nisbet extends the history of progress to Hesiod of the late eighth century B.C. (Hull, in this volume argues otherwise). I believe that progress is now, and always has been associated with change, development, direction or trend. Nisbet may be right that the Enlightenment's idea of progress had a meaning not much different from that of the Ancients. The Enlightenment's stages, like the paleontologists' epochs and periods, in which succeeding periods are improvements on the preceding, were similar to the ideas of social and intellectual progress in Hippocratic writers and in Thucydides, and to single directions of the Ancients. This led the Ancients just as well as the eighteenth century Aufklärung to the next evolutionary stage: the recognition of progress of various existing cultures where "primitiveness" and "progressiveness" were recognized.

Nevertheless, most historians follow Bury (1932), who divided the history of the idea of progress into three periods. The first, concerned with philosophical and historical examinations, ended with the French Revolution. The second, associated with "development," consisted of a search for laws of progress and of attempts to establish it. The third began with the publication of *The Origin of Species* in 1859. It is in this last period that evolution replaced providence as a causal agent of progress. Bury, like most other historians, emphasized that by itself evolution is neither optimistic nor pessimistic, but to base progress on evolution it must be shown that human social life progresses toward improved reason and improved moral conditions and thus towards greater happiness. Bury, again like many other historians, depended on Darwin's famous passage that "as natural selection works solely by and for the good of each being, all corporeal and mental endowments will tend to progress towards perfection" (Darwin 1859, p. 669).

In the present volume, papers by Robert J. Richards and by Robert C. Richardson and Thomas C. Kane are concerned with the nineteenth century: Spencer, Darwin and the neo-Lamarckians. Richards, in his studies of the history of evolution, argues that Darwin's ideas were similar to Spencer's and that Darwin, like Spencer, based his notions on theological and moral foundations interpreted in terms of their time. To Richards, Herbert Spencer is not a man who saw the moral and social "nastiness" formed around the "survival of the fittest" view. On the contrary, Spencer emerges as an advocate of a perfect future world of freedom, happiness, and equality for all in which "progress is . . . not an accident, but a necessity," a Spencerian phrase so frequently quoted by students of Spencer. Spencer's Law of Organic Progress was the law of all progress -- of evolution unfolding. According to Richards, Spencer's theological concept of human goals was based on his early religious beliefs and was later framed in natural laws designed to produce evolutionary progress.

It is mostly in the interpretation of Darwin's idea of progress that Richards differs from the mainstream. Richards very convincingly and abundantly documents that Darwin's views of progress were indeed very similar to Spencer's. He shows this in detail in Darwin's early letters, in his notebooks, and of course in *The Origin*. Richard's conclusion is inescapable: Darwin certainly believed in progress, in perfection of life, in evolution being progressive (but cautiously handled), in intellect and moral sense being obvious ideals of perfection which infuse nature with a higher purpose, and in progress being "not a result of some innate tendency or inner force toward perfection" but "an external dynamic condition." The genius of Darwin was in weaving together the biological evolutionary concepts of progress with a strong moral progress, and in his arguments that natural selection is responsible for progress and the moral improvement of man.

After scrupulously analyzing the ethics and philosophy of Spencer and of Darwin and their concepts of progress, Richards concludes that, to both, progress made sense above all as a *moral* development.

The second article in this section is by Richardson and Kane, who are interested in the problem of progress as debated by late nineteenth century evolutionary theorists. Richardson and Kane are primarily concerned with the genesis of American evolutionary thought, particularly the role of orthogenetic mechanisms in the work of early neo-Lamarckians. Their secondary interests are in an analysis of cave faunas as a paradigm for the neo-Lamarckian program.

Richardson and Kane offer a balanced survey of the nineteenth-century American neo-Lamarckian school. The best-known representatives of this school, Alpheus Hyatt and Edward Cope, promoted a modified form of Lamarckism. Cope concentrated on the effects of use as an adaptive agent of evolutionary change, and traced the evolutionary trends he saw linking various fossil specimens. Hyatt admitted that evolutionary developments often involved utilitarian modifications, but he also found the pattern in the fossil record incompatible with Darwinian assumptions. While a variety of patterns were realized (trends to increased specialization and increasing complexity), Cope construed the patterns as progressive, while Hyatt did not. Nevertheless, although both writers recognized trends, neither required progress to have *necessarily* occurred. Evolutionary lineages tend to be modified in different directions; therefore, there could not be any inherent or necessary direction to evolutionary changes.

We are all aware that there were two Darwins, one of *The Origin,* with its gradual modifications produced by natural selection, and the other of *The Variation of Plants and Animals under Domestication,* with its hereditary mechanism of pangenesis. What Richards forces us to see is yet two other Darwins, one of the biology of evolution, and the other of the philosophy of evolution. Richards has peeled off the evolutionary stratum to examine in detail Darwin's metaphysics. The result is a surprise to those of us who regard Darwin only in the classical evolutionary light. What Richardson and Kane have done, on the other hand, is to examine the Darwinism of natural selection. These chapters put Darwin's views in complementary perspectives. Richards presents Darwin as a progressionist; Richardson and Kane's Darwin *may* be an evolutionary progressive but one who nevertheless continuously and cautiously insists that his natural selection need not include progressive development. Richardson and Kane interpret the interaction of Darwin and his orthogenetic critics as an argument *not* concerned with progress, but with the *mechanism* of evolution. Thus while orthogenetic interpretation of evolution may have demonstrated some linearity in evolutionary processes, this linearity *did not* require progress.

What is perhaps novel in Richardson and Kane's analysis of Hyatt and Cope is that neither Hyatt nor Cope, the main proponents of orthogenesis, advocated evolutionary progress, yet at the same time they recognized evolutionary trends. The neo-Lamarckian school claimed that evolutionary changes were basically of orthogenetic character and that these changes were nonadaptive. Cope argued that trends in the fossil record tended to be progressive, while Hyatt thought them to have no necessary direction. Both, however, urged that evolutionary development depended upon non-Darwinian factors.

In the previous section, Ayala dissected the idea of progress, and in the next two chapters his definitions become applicable in the analysis of different sorts of sociocultural evolution. That progress is real can be demonstrated, and actually even measured in one very important sense -- in characters that are acquired and somehow inherited. One must immediately explain, however, the term *inherited*. Surely language, behavior and all culture are inherited -- but not in the sense that morphology or physiology is. Tools and culture are not genes! They are, however, passed to the next generation and they become a permanent component of the society, as well as the causes of the cultural evolution of man. These are the main thrusts of the chapter by anthropologist Robert C. Dunnell.

Dunnell examines the idea of progress in the context of cultural anthropology and archaeology. The notion of progress, he says, has been the central focus of cultural evolution, and cultural anthropologists who are not biologists often apply biological terms in a different sense from biologists. Darwinian evolution is not applicable to cultural evolution because the mechanisms of inheritance are different for biological and for cultural traits. As a consequence, cultural evolution has been frequently confused with biological evolution. Dunnell ably argues that cultural evolution either directly, or as a methodological model, rarely uses or applies the theory of biological evolution. The main difference he recognizes lies in the concept of progress which for biologists is an observed feature of evolutionary change, while for cultural evolutionists progress also causes cultural change.

This very important point leads to his startling conclusion that the "theory" of cultural evolution is identical with the empirical claims of cultural evolution -- and hence an untestable redundancy. But attempts have been and are being made to develop mechanisms that would allow the application of the concepts of biological evolution to cultural phenomena. These attempts, however serious, are still nonoperational, and hence the Darwinian interpretation is still not widely acceptable among anthropologists and archaeologists. Nevertheless, Dunnell predicts that eventually cultural progress will be explicable in Darwinian terms.

How is the idea of progress exemplified in the Socialist Countries, particularly in the USSR, where the dialectic interpretation of biology is favored and where faith in human abilities to improve the future is still strong? To a Russian, purpose *need* not imply any force (divine or otherwise), and Marxism more than any other philosophy seeks to understand history and the social processes responsible for historical changes. The Russians see progress in life as a stochastic process caused by the blind forces of natural selection. In the West, on the other hand, because progress implies purposefulness, biological progress is rejected, simply because God and purpose are rejected. In the Soviet model there is no trace of purposefulness. And it is this that concerns Adam Urbanek.

Urbanek examines the theory of morphological progress as formulated by the Soviet biologists, particularly the comparative anatomist and evolutionary biologist Alexei Severtsov, and his followers. Severtsov's concept of aromorphosis constitutes the core of the subsequent ideas of evolutionary progress representing a great advance toward a better understanding of a number of theoretical notions. Severtsov demonstrated that evolutionary changes cannot be measured merely by the biological success of population, species or lineages. He suggested that evolution involves also an entirely new dimension -- the improvement in the morphophysiological organization of the body. Severtsov characterized the events leading to such improvements as a distinct class of evolutionary changes. Thus, Soviet biologists stress the profound analogy of biological progress with technical and social progress -- and this represents the most ambitious attempt to combine the concept of progress in biological and social spheres into a single theoretical system. Urbanek had a difficult task of compressing such a varied wealth of Soviet biology into the seemingly rather dry determinism of present-day evolutionary ideas.

Severtsov's views on evolution appear to have had a great influence on

Soviet biologists and paleontologists. It appears that the Soviet controversy is similar to our microevolution/macroevolution debate, particularly experimentalist versus anatomist. Is aromorphosis a macroevolution and is idioadaptation a microevolution? Here is Urbanek's great contribution: There is very little communication between Soviet and Western biology, and Urbanek has brought the significant evolutionary Soviet view closer to us.

Empirical Approaches

How is progress seen in evolutionary biology, and how is it expressed by its practitioners? To J. Marvin Weller (1969), the whole field of paleontology is impregnated with the concept of progress. It is never a question of destiny but a question of betterment. Progress is an essential element of all evolution and is present in the entire history of life (1969, p. 20). Anything that becomes better adapted and more efficient is a document of progress.

To D'Arcy Thompson (1969), on the other hand, evolution from simpler to more complex, so dear to Weller, is no evidence of progress. "That things not only alter but improve is an article of faith, and the boldest of evolutionary conceptions . . . . I for one imagine that a pterodactyl flew no less well than does an albatross, and that Old Red Sandstone fishes swam as well and easily as the fishes of our own seas" (1969, p. 201).

George Gaylord Simpson (1947) understands natural selection as "the positive and creative process" (p. 207), rather than, or in addition to, its negative function as the great eliminator of undesirable traits. According to Simpson, it "achieves the aspect of purpose without the intervention of a purposer" (p. 212). He thus believes natural selection to be a creative process, proven both theoretically and experimentally.

Sir Julian Huxley, an illustrious English evolutionary biologist and a United Nations administrator, imagines that man must acknowledge "that he is the highest entity of which he has any knowledge" (1941, p. vi). Huxley traces the progress of man from the extremely slow inorganic sector, through the tempo-quickened organic or biologic sector, to the quickest psychologic or psychosocial human sector (1960). As Huxley understands the process, the self-reproducing matter conceives a biological phase to beget a self-reproducing mind to give birth to man. One sector succeeds another in time, and the later evolves from the earlier. All this inevitably and surely leads to man (1960, p. 19). In the psycho-social evolution the selective mechanism is a goal-selective mechanism. Man's aim is the fullest realization and fulfillment of more possibilities by the human species collectively.

A more contemporary look at these issues is presented by the last five papers in this section of the book. No writer has influenced contemporary evolutionary genetics more than John Maynard Smith. In a very concise and lucid paper this eminent researcher argues that biology denies progress but that life nevertheless goes through "advancing" sequences. Progress, as already mentioned, is slighted by biologists for semantic, anthropomorphic and other reasons. It is a very difficult idea to define (though see Ayala's contribution); it is often assumed to have led to man; it is frequently inferred from the ladder of life, or the great chain of being; it is repeatedly interpreted in terms of morphological complexity either in terms of structure or amount of the coding DNA. But the *"theory of evolution does not predict an increase in anything,"* and Maynard Smith, through brilliant examples, demonstrates that evolution is different from other physical processes; it seems that evolutionary changes are without time's arrow. But above all there is very little empirical reason to support the proposition that evolutionary processes are progressive. Maynard Smith is not concerned with the fossil record (for discussion of the fossil record see the chapters by Raup and Gould) and neither does he see progress in "evolution in vitro," for example, in the replication of RNA molecules.

Maynard Smith describes the sequences and the levels of complexities through which life had to pass. Within the Darwinian theory of evolution these must have occurred because living entities exhibit multiplication, variation and heredity. According to Maynard Smith, the "units of evolution" are genes, genomes, organisms, demes, species and communities. These units have passed through increased levels of complexity. How it is that selection at one level does not disrupt integration at the next higher level is answered by Maynard Smith, who concludes that although progress need not be an inevitable result of evolution, nevertheless there are real "revolutions in the way in which genetic information is organized." The disruptive effects of selection at lower levels have been suppressed. And this is an important common feature of all units of evolution.

William C. Wimsatt and his student Jeffrey C. Schank present evolutionary change on a macroevolutionary time scale through their investigations of developmental and selective constraints on the evolution of complex adaptations. They propose four connected mechanisms that allow the building up of more complex adaptations than would otherwise be possible. They propose a new constraint: that the genes contributing to a complex adaptation of a unit of selection must show differences in the size of their fitness contributions. Wimsatt and Schank show how this constraint differs from "Darwin's principles," which require heritable variance in fitness. They argue that it is exceedingly robust and is favored independently by basic physical considerations and by selection.

Discussion of genetic load suggests that there is a maximum number of genes that can be maintained by selection. Wimsatt and Schank propose four mechanisms through which, over extended periods of evolutionary time, this number can and should significantly increase by two to five orders of magnitude.

The authors argue that reduction in the average amount of genetic load per locus should allow genes to be maintained by selection at more loci, and thus a larger genome can be maintained by selection. The first three mechanisms suggest long range changes, not only in the size but also in the architecture of the genome and in its mode of expression. These mechanisms suggest that progressive increases over evolutionary time in the maximum size and complexity of developmental programs are virtually inevitable. Their simulations, however, have a broader significance: constraints are creative and progressive in channeling and directing variations. "This is a deep truth, not only about evolution, but about problem-solving and exploration in general." They follow Simpson in seeing natural selection as a creative force rather than as only a great eliminator. This paper represents the first new theoretical argument for a kind of evolutionary progress, based on population genetic and microevolutionary considerations that could signal a new turn in evolutionary discussion of progress.

E. O. Wiley has been concerned for some time with nonequilibrium thermodynamics and evolutionary processes. Here he probes the very difficult subject of entropy and progress. There were other earlier workers (e.g., Blum

[1951], who claimed that the second law is applicable only to isolated systems that are left to themselves and thus will run down; and that life is not isolated) who followed the initial suggestion of Eddington (1935) that all life follows the second law of thermodynamics.

Wiley builds on Brooks's and his recent book on evolution and entropy (1986) and proposes that evolution is a special case of the second law. Biological systems such as organisms and higher taxa exhibit irreversible behavior that is decidedly different from physical systems. Wiley believes that the fundamental difference between living and nonliving is in the organism's "instructional information." That is, only biological organisms, not the nonliving entities, possess information that actually resides within the organism itself (e.g., DNA molecule) -- and not outside the system in the environment. Wiley imaginatively demonstrates with analogies how organisms carry their blueprints internally.

But are these properties unique to the living? Do not individual atoms also carry their blueprints internally? Is not the disintegration of radioactive elements controlled by "instructional information"? Is the "chemical behavior" of atoms not internally controlled? Wiley postulates that organisms obey all the thermodynamic rules prescribed to all nonliving structures, and above all that the path of evolution is expressible in terms of an entropy function which measures order, complexity and organization at all hierarchical levels. According to Wiley, evolution is a nonequilibrium process. He concludes that the arrow of evolution is in the direction of increasing entropy and increasing organization. However, this does not imply evolutionary progress but rather historical constraints on species. These constraints may make evolution appear teleological, but evolution is neither teleological nor orthogenetic.

To establish that progress has indeed happened a certain value judgment must be identified, and the stage at which progress occurred must be shown to be better than the previous stage. There is little reason to object to the biological identification of the term "better." After all, we do identify what is necessary for an organism to sustain and prolong life and what is detrimental. Once we recognize this, the "problem" of better and worse disappears. Therefore we should be able to identify such values if they occur in the geological past. This is what David M. Raup has accomplished, postulating that what prolongs the life of a taxon may be good and measurable, and hence the evidence of progress in the history of life could be theoretically determined. Raup has pioneered the study of the patterns of diversity, particularly their manifestation in the fossil record, and is in the unique position of having something that can be actually measured. The fossil record, in spite of claims to the contrary, is sufficiently good and the overall pattern that emerges from the ranges of at least a quarter million species probably will not change.

Raup shows clearly the difficulties and pitfalls of how and what parts of the fossil record should and could be measured. Raup and Sepkoski (1982), just like Van Valen (1984), have found that throughout the Phanerozoic there is a strong tendency for taxonomic groups to exhibit an increased resistance to extinction. Raup, however, is very careful about drawing conclusions from this. Although his analysis of the paleontological data shows that indeed there is an increase in taxonomic survivorship through the Phanerozoic, he points out a number of reasons why his conclusions are tentative and not yet unequivocal and why they need further scrutiny.

To Stephen Jay Gould, as to many others in the present volume, progress is a result of cultural conditioning only. However, new evolutionary data allow him to reinterpret the concept of progress. He advocates the abandonment of the idea of progress, and proposes to install in its place a new, very real, deep and essentially directional pattern. To him the history of life has a directional change, whose directionality he examines. Gould summarizes the history of the idea of evolutionary progress, as seen by working evolutionists. His discussion of real trends in clades, faunal replacements and biomechanical improvements will always be "good reading" -- but his critiques of these concepts will became "required readings."

Gould sees that the "new evolution" (that not all of us are capable of seeing) will reinterpret all documented trends according to the hierarchical principle of interacting levels. Trends are differential sortings of species and not gradual transformations of populations. Most premises about progress were made *a priori* for the *ipso facto* conclusions drawn in favor of improvement.

But above all trends are changes in variance -- just examine Jablonski's (1987) interpretation of Cope's rule, the result of increase in variance and not a directional change in body size. "Monads to man" is no evidence for progress; monads could not attain negative sizes, therefore their only path was toward increased size and complexity. Gould's own "new evolution" begot exaptation (a useful feature originated for a different reason) which, rather than adaptation, is responsible for progress.

But there are requirements for history to become a history. The historical events must have temporal coordinates -- or history will not be sequential. History must also have directionality, or any asymmetry by which the time's arrow can be measured. Gould's bottom-heavy clades are just one such example (see Gould et al. 1987). The directionality with which Gould replaces progress has been discussed by Lotka (1945), and by Williams (1966 and references therein), and is implicit also here in the papers by Hull, Raup and Wimsatt and Schank.

Progress as a Semantic Problem

In this last quarter of the twentieth century our interpretations of evolutionary events are highly speculative and highly statistical (due to the work of Sewall Wright and perhaps the sprouting of computers?), and a high value is set on the search for synthesis (however openly admitted). In the nineteenth century this synthesis was embodied (however consciously) in progress. Now as the new data claim to shake the old cathedrals of evolutionary theory, "better" and "newer" theories require the establishment of progress to be more rigorous. What is demanded is the demonstration of specific and not general or universal progress. Above all, *direction replaces progress*.

William Provine attempts to show the deficiencies in the synthetic theory and in the concept of progress. David Hull is most skeptical of all evolutionary direction and, contrary to Robert Richards, does not believe that Darwin saw progress in evolution. Michael Ruse, on the other hand, sees progress in one form or another in all of us. Francisco Ayala clearly differentiates between universal and limited progress. Surely universal progress is for all times and for all situations, hence it can exist only when its aim is infinite. Progress ceases to be universal when its aim has been reached. It is very difficult to visualize a universal progress, and therefore we can consider only "limited" progress -- which all the contributors to the present volume accept. Robert Dunnell sees progress on the one side as

historical order, trends and direction, and on the other side progress assumes
the role of a theoretical tenet --which explains the historical events as caused
by progress -- more than a dangerous consequence. New meanings of progress
are proposed. John Maynard Smith possibly recognizes these in levels of
selection (and, to him, for progress to be demonstrated, it need not have
occurred in all branches of the tree of life, but, more plausibly, was
sufficient in at least one); to Bill Wimsatt and Jeffrey Schank it is a
consequence of two constraints on complex adaptations and the mechanisms
which channel their effects; to Ed Wiley the direction of evolution is in the
direction of increasing entropy; David Raup may perceive progress as an
increased resistance to extinction; and to Stephen Gould it is only obvious in
the directionality in history and in the bottom-heavy clades. Gould, just as
Hull, strongly proposes that our scientific activities are culturally controlled.
Perhaps Gould goes further than most of us: he is known to ascribe much of
our personalities and intellectual characteristics to environmental causes. But
can we not, if Gould is right, also claim that our acceptance or rejection of
evolution and its associated progress is also culturally determined?

Without these new meanings of progress a new synthesis will be
impossible. Either a new synthesis will be found, in which case new meanings
will became new philosophy, or "no synthesis, no philosophy." For example,
"new catastrophism" assumes non-Darwinian processes and this process becomes
its philosophy -- the new philosophy becomes the subject and product of the
new scientific theories.

We might consider the role of progress in the construction of models in
evolutionary biology. In order to arrange morphologies we must have a model,
with a certain direction of change in morphology that can be explainable and
chronologically arrangeable. These models are further refined since the
general and broad models are insufficient. That is why progress is a useful
concept. But we must always remember that models *only evoke* the reality of
morphological changes. The continuous change of morphology, I suspect,
cannot be grasped any more than the concept of time can be grasped. Models
must never be taken too seriously. Any model, *all* models, are just
descriptions of many other possible models. This is not to say that models are
confused or should not be made. Although *comparison, dit le proverbe, n'est
pas toujour raison*, we are always building models, and we are always making

comparisons because they seem best suited to explaining complex processes. Progress may also be best explained perhaps by comparing it with some technological innovations. Consider the development of transportation as a model. If we define transportation (without moral or ethical values) as a process of moving certain mass over a given distance in a specific time -- then there exists an undoubted progression in speed, distance, and mass moved from the "primitive" means of pulling weight by hand on land to the great diversity of land, water, air and space taxa with an incredible array of means of motive power. No one can deny the existence of progress from a single pony-drawn carriage to the supersonic jet. (Note that extinctions or near extinctions abound along the way and are always associated with explosive innovations.) But here we have an example of progress that became universalized. This progress can be objectively measured. It is interesting that since we defined progress in terms of mass, space and time, even when value judgments are superimposed, the increased speed, mass and distance show progress to be general. Even the most ardent pacifists will accept that the military use of transportation did not diminish the universal value, or "progress," in transportation. Neither does the assertion that the only way for transportation to proceed is by increasing its complexity detract from the reality of progress in this case. We have thus demonstrated that *improvement* in means of transportation is progress irrespective of its utility and its subjective values.

How does our example apply to evolutionary biology? A taxon is a group of organisms with a common morphology. It is defined when this morphology is defined. When speciation occurs (or better still when "taxonation" occurs), this morphology (or form) increases its spread. The process of taxonation can be seen as a spread of genes. When this form is selected against, extinction of form (or morphology) occurs. Thus the form or a certain morphology can spread or become extinct, can be associated with an increase of complexity, or be considered progress. However, diversity can increase with or without a concurrent increase of complexity. Often a certain morphology is identified in the fossil record, and then it is assigned a degree of primitiveness or advancement. However, these morphological concepts are abstractions taken out from the totality of the fossil record. They do not possess reality of their own and only exist as ideas. When these morphologies are arranged in series, a direction appears. Large and sudden changes in morphologies are recognized as extinctions or "adaptive" radiations. Yet the totality of morphology and morphological changes are probably never fully grasped. It seems sufficient at this stage to refer to the realm of limited progress as suggested in three recently published studies of the fossil record. One is by the Field Museum paleontologist Scott Lidgard (1985, 1986), who shows an evolutionary trend in growth patterns of cheilostome bryozoans consistent with a persistent evolutionary direction toward an increased colony integration. This he capably shows for both fossil and recent forms (for further discussion see the chapter by Gould). Another is by Frank K. McKinney (1986) who elegantly demonstrates just this kind of progress in the fossil bryozoans as a whole. McKinney showed that the branched "unilaminate" bryozoans during the era-scale length of time increased their diversity by a factor of ten relative to other colony forms, in which no significant trends are recognized. These filter-feeding bryozoans offer the least resistance to the outflow of water and are thus distinctly "better" adapted than any other erect bryozoans. Lastly, Alan Cheetham synthesizes bryozoan colony growth, biomechanics, and fossil history to demonstrate an evolutionary increase in the resistance of erect colonies to failure in flow, a trend mediated by changing skeletal design rather than material strength, and most importantly one that seemingly arose through "selection among species of generally advantageous morphological differences that are species-specific and fluctuate randomly within species" (1986, p. 151).

However, we have *not* demonstrated that progress is universal, only that it is limited to certain specific areas of human activities and to selected cases in the evolution of fossil invertebrates. In order to show the total universal progress in human history and in the nonhuman biological realm, we must show that the sum of progress in the totality of all specific areas is greater that the sum of all regressions in human (and nonhuman) history. This has not been done, and our volume demonstrates that it cannot be done. What remains is progress in statistical terms of short durations, limited scopes, and in some of the many dimensions available for the description of evolutionary change. Some of these perspectives may provide useful grist for the mill of future evolutionary debates, for where the concept of progress is concerned, that there will be continuous debate is perhaps the only certain conclusion. Or whether or not we downplay its importance, or even downright deny it, as

Michael Ruse claims we do, are the thoughts of progress not present in us all?

References

Bergson, H. [1907. *L'Évolution créatrice*.] 1944. Creative Evolution. Translated
    by A. Mitchell. New York: Random House.

Blum, H. F. 1951. *Time's arrow and evolution*. 3d ed. 1968. Princeton: Princeton University
    Press.

Brooks, D. R., and E. O. Wiley. 1986. *Evolution as entropy. Toward a unified theory of biology*.
    Chicago: University of Chicago Press.

Bury, J. B. [1932] 1955. *The idea of progress. An inquiry into its origin and growth*.
    Reprint. New York: Dover.

Cheetham, A. H. 1986. Branching, biomechanics and bryozoan evolution. *Proceedings of the
    Royal Society of London. B. Biological Sciences* 228:151-71.

Darwin, C. 1859. *The origin of species*. 6th edition. 1872. London: John Murray.

Eddington, A. 1935. *The nature of the physical world*. London: Dent & Sons.

Gould, S. J., N. L. Gilinsky, and R. Z. German. 1987. Asymmetry of lineages and the
    direction of evolutionary time. *Science* 236:1437-41.

Huxley, J. S. 1941. *Religion without revelation*. London: Watts & Co.

Huxley, J. S. 1960. The emergence of Darwinism. In *Evolution after Darwin*, ed. S. Tax, 1-
    21. Vol. 1. Chicago: University of Chicago Press.

Jablonski, D. 1987. How pervasive is Cope's Rule? A test using Late Cretaceous mollusks.
    Geological Society of America. Abstracts with Programs 19:713-14.

Lidgard, S. 1985. Zooid and colony growth in encrusting cheilostome bryozoans.
    Palaeontology 28:255-91.

Lidgard, S. 1986. Ontogeny in animal colonies: A persistent trend in the bryozoan fossil
    record. *Science* 232:230-32.

Lotka, A. J. 1945. The law of evolution as a maximal principle. *Human Biology* 17:167-194.

McKinney, F. K. 1986. Evolution of erect marine bryozoan faunas: Repeated success of
    unilaminate species. *The American Naturalist* 128:795-809.

Nisbet, R. 1980. *History of the idea of progress*. New York: Basic Books.

Raup, D. M., and J. J. Sepkoski, Jr. 1982. Mass extinctions in the marine fossil record.
    Science 215:1501-03.

Ruse, M. 1986. *Taking Darwin seriously: A naturalistic approach to philosophy*. Oxford:
    Blackwell.

Russell, B. 1927. *Selected papers of Bertrand Russell*. New York: Random House.

Russell, B. 1935. *Religion and science*. London: Oxford University Press.

Russell, B. [1944] 1963. *The philosophy of Bertrand Russell*. Vol. II, ed. P. A. Schilpp.
    Harper Torchbook Edition. New York: Harper & Row.

Russell, B. 1945. *A history of western philosophy*. New York: Simon and Schuster.

Simpson, G. G. 1947. *This view of life. The world of an evolutionist*. New York: Harcourt,
    Brace & World.

Thompson, D. W. 1969. *On growth and form.* Abridged edition, 1961. Cambridge: Cambridge University Press.

Van Valen, L. 1984. A resetting of Phanerozoic community evolution. *Nature* 307:50-52.

Weller, J. M. 1969. *The course of evolution.* New York: McGraw-Hill.

Williams, G. C. 1966. *Adaptation and natural selection: A critique of some current evolutionary thought.* Princeton: Princeton University Press.

PHILOSOPHY OF PROGRESS

Progress in Ideas of Progress

David L. Hull

According to the traditional view, no one throughout human history found the idea of progress either in nature or in the course of human affairs plausible or appealing until the Renaissance. The ancient Greeks and later Romans viewed the world in terms of eternal cycles, while Christian theology portrayed human history as a period of tribulation between Adam's fall and the Second Coming. Not until the sixteenth century did intellectuals in the West begin to think that possibly human history as well as nature at large might be progressive, culminating in *la bellé epoque* of the idea of progress between 1880 and 1914. During this period, the belief that everything in every way is getting better and better was pandemic. World War I brought this intellectual idyll to a close (Bury 1932). The Great Depression and World War II did not do much to rekindle the hope that all is for the best. Even Wagar's (1972) assiduous search for some glimmers of a thaw in the glacial spirit that characterized the period after this great divide results in only the most tenuous hints of a belief in progress. Then in the 1960s, everything began to change:

> Evidences of revival are everywhere: a new "secular" theology of hope, the emergence of new radical ideologies in the West and a more humanistic socialism in Eastern Europe, the neo-Marxist and neo-Freudian thought of Fromm and Marcuse, the posthumous fame of Teilhard de Chardin, the return to fashion of evolutionism and theories of change and progress in the social sciences, and the new charismatic politics of the Kennedy brothers, Martin Luther King, Malcolm X, de Gaulle, Castro, and Che Guevara. (Wagar 1972, p. 243)

The Age of Aquarius was upon us.

Revisionist history is currently popular, and the idea of progress has not

27

escaped the relentless grinding of its insatiable gizzard. According to Nisbet (1980), traditional historians got the early story of the idea of progress all wrong. Cycles and the fall of Adam notwithstanding, the idea of progress began with the Greeks and has continued unabated through the Romans and the Christian fathers until the end of the Middle Ages when it suffered an eclipse during the Renaissance. The Reformation reintroduced a belief in both slow, gradual, cumulative improvement in knowledge and the goal of moral and spiritual progress. Thereafter, Nisbet's chronicle coincides with the traditional story until the present. Anyone who surveys current literature, art, philosophy, theology, scholarship, and science is likely to agree with Nisbet (1980, p. 318) that among intellectuals "disbelief, doubt, disillusionment, and despair have taken over," but this message seems not to have gotten through to the tens of millions of people in the West who maintain an indestructible faith in progress even as Western culture declines dramatically.

Like Wagar, Nisbet (1980, p. 356) sees some hope for a revival of a belief in progress, possibly even a revival of progress itself; but his hope is kindled not by radical ideologies, the Kennedy brothers, or Che Guevara but by the "faint, possibly illusory, signs of the beginning of a religious renewal in Western civilization, notably in America. Whatever their future, the signs are present -- visible in the currents of fundamentalism, pentecostalism, even millennialism found in certain sectors of Judaism and Christianity." Nisbet agrees with Wagar in at least one respect, that Teilhard de Chardin will hold a prominent place in a fusion of science and religion "based on the inexorable progress of human knowledge" that Nisbet (1980, p. 316) hopes will come to pass. I cannot speak for Nisbet's tens of millions of ordinary people, but the likes of Jerry Falwell, Pat Robertson, John Paul II, and even the kindly Teilhard instill very little confidence in me about the future. Perhaps overpopulation, pollution, the greenhouse effect, the depletion of the ozone layer, AIDS, and the continued persecution by governments and organized religions alike of those who are most vulnerable are all illusions or minor inconveniences, but I do not think so.[1]

A popular view at present is that scientific beliefs are not as immune to the rest of culture as the evil positivists once were supposed to have claimed. The history of the rise and fall in the popularity of the idea of progress among scientists in the West provides an excellent example of the interpenetration of science and the rest of society. In biology the issue was the relative superiority of one organism to another. The conception of living creatures forming a Great Chain of Being has been and continues to be one of the major themes of Western thought (Lovejoy 1936). Of course, throughout most of human history, this great chain was considered to be as timeless as the periodic table of physical elements is considered today. Aristotelians ordered species from the simplest plants through corals and worms to cephalopods and human beings. They did not think that species arose through time in this order or that organisms progressed through successive generations along this chain. Early attempts to "temporalize" the Great Chain of Being left the timelessness of this chain untouched. For example, in later years Linnaeus believed that only a relatively few species were created in the beginning and all subsequent species arose through the crossing of these primary species. Thus, secondary, tertiary, etc., species arose through time but according to a timeless plan set down when the original primary species were created. Linnaeus's primary species established taxonomic space as surely and timelessly as the atomic structure of the physical elements determines the number and kinds of possible compounds which can result from their combination. Some of these kinds may never be exemplified, but they are built into the structure of the universe just as surely as those that are. For most of the history of biology, species were considered to be in this sense eternal and immutable.

Contrary to popular belief, Lamarck left the living world almost as static and timeless as he found it (Simpson 1961a; Hull 1967; Gould 1977). According to Lamarck, simple organisms are generated spontaneously, albeit mechanically, out of inorganic material and then proceed to progress up one of several trees of life under the impetus of an innate though mechanical urge to perfection. But mirroring these progressive trees of life, Lamarck postulated sequences of degradation that reduce organisms belonging to these more complicated species to simpler inorganic products. Organisms might change their species, but these species themselves did not change. They did not "evolve" in the modern sense or even become extinct, at least not for long. Instead Lamarck considered organic and inorganic species to be equally part of the framework of the universe. Lamarck's world of entities cycling endlessly through timeless categories was nearly as eternal, immutable, and cyclical as that of the early

Greeks. If an alchemist's transmutation of a few ounces of lead into gold counts as "evolution," then nearly everyone from Aristotle to Agassiz were "evolutionists." I prefer a terminology that makes at least a few basic distinctions.

Creationists and ideal morphologists alike viewed the fossil record as revealing clear signs of progress, disagreements about mechanisms to one side. Lyell and Huxley did not share this perception. Their interpretation of the fossil record indicated no direction, let alone progress, only a steady state. Darwin's views on progress are a matter of some controversy. The traditional view is that Darwin had his doubts about biological evolution being progressive. It might as a contingent fact have a direction of sorts, e.g., increased complexity, but it did not exhibit any innate tendency to progressive development (Darwin 1903, 1:344). Richards (this volume) disagrees. Although he finds Darwin handling the idea of progressive evolution cautiously, he interprets Darwin as viewing natural selection as an "organ to manufacture biological progress and moral perfection." As in the case of Herbert Spencer, Darwin's theory was "powered by a law" that would produce the "kind of evolutionary progress that their theological and ethical sentiments demanded." As much as natural selection operated by means of death and destruction, on Richards's reading, it had a "moral heart" that made progress a general, though not an invariable, rule.

What I find remarkable about Darwin was that at a time when a belief in progress was pandemic, he had so little to say about it and, when he did, expressed himself so equivocally. In all 490 pages of the *Origin of Species*, he mentions progress only a dozen times, and in half of these cases he means only change. As in the work of previous authors, two questions were at issue for Darwin: does the fossil record indicate some sort of progress, and does the theory one holds about the genesis of this record imply that it should be progressive? On both counts Darwin expressed doubts. For example, he asked which are "higher," molluscs or crustaceans? According to Darwin (1859, p. 337), crustaceans are not the highest forms in their own class, yet they seem to have beaten the highest molluscs in the struggle for existence. Mollusca includes such simple forms as snails and clams as well as such higher forms as cephalopods and species of extinct nautiluses measuring thirty feet across. Crustacea includes water fleas and Darwin's own barnacles, but it also includes such complex forms as lobsters, and the closely related trilobites are among the most numerous extinct forms.

On the basis of his theory, Darwin reasoned that later fauna and flora under nearly similar conditions should exterminate earlier forms the way that extant species from Great Britain were replacing native forms in New Zealand. "I do not doubt," Darwin (1859, p. 337) wrote, "that this process of improvement has affected in a marked and sensible manner the organization of the more recent and victorious forms of life, in comparison with the ancient and beaten forms." But he quickly added, "I can see no way of testing this sort of progress." Regardless of Darwin's thoughts on the subject, his contemporaries found the survival of the fittest to be anything but heartwarming and reassuring. Death, waste, extinction, and misery seemed to be the driving force of Darwinian evolution.

Although the few pronouncements that Darwin made on the idea of progress allow room for disagreement about his changing views on the subject, there is no doubt with respect to his ideas about the other characteristics of the evolutionary process. For Darwin evolution was largely gradual, and natural selection was its chief directive force. Yet, the Darwinian theory that swept across Europe like the Flood, treated speciation as saltative, included natural selection as a subsidiary mechanism, and took for granted that biological evolution is progressive.

One often hears, in support of broader cultural influences on the substantive content of science, that Darwin's theory became so popular after 1859 because Darwin read British capitalism into the biological world. Yet, the versions of evolutionary theory that gained a wide following among scientists at the time (not to mention the general public) had none of the characteristics of British capitalism. *Zeitgeist* might well have influenced the reception of evolutionary theory among scientists and the public at large, but its influence seems to have been to transform Darwin's theory into something more palatable. Instead of unlicensed competition, Darwinism came to imply endless improvement. Perhaps narrowly scientific considerations played a role in the transformation of Darwin's theory, but a predilection for progress also seems to have been operative (Glick 1974).

Anthropocentrism

One of the great discoveries of the ancient Greeks was that man himself is the wonder of the world. Such anthropocentrism has characterized Western thought ever since. Needless to say, early evolutionists considered the human species to be the ultimate goal of biological evolution. For example, Wallace (1889, p. 47) can be found saying that:

> To us, the whole purpose, the only *raison d'être* of the living world --
> with all its complexities of physical structure, with its grand kingdoms,
> and the ultimate appearance of man -- was the development of the human
> spirit in association with the human body.

As hardheaded as Darwin (1859, p. 489) attempted to portray himself, even he included in his poetic conclusion to the *Origin of Species* the remark that "as natural selection works solely by and for the good of each being, all corporeal and mental endowments will tend to progress towards perfection." Although such blatant anthropocentrism is currently out of fashion, it continues to crop up in the most unlikely places, for example, in the classic text on evolution by Dobzhansky et al. (1977, pp. 513-16), albeit apologetically (see also Ayala this volume). After several explicitly stated reservations, these authors note that a variety of criteria can be used to order kinds of organisms into progressive series. As an instance, they cite increased ability to gather and process information. Bacteria do not make the grade at all, paramecia just barely, and euglenae somewhat more. Even the most complex plants are inferior to the lowliest animals in their ability to obtain and process information. Sponges have no nervous systems at all, but coelenterates possess at least an undifferentiated nerve net. Starfish exhibit some progress over coelenterates, but planarian flatworms are the first organisms to possess a rudimentary brain. Insects come next in the progression followed by vertebrates, mammals, and finally the crowning glory -- *Homo sapiens*. The suspicion that this sequence of biological taxa has only the most tenuous connection to phylogeny is heightened by noting that Aristotle could have made exactly the same observations.

Regardless of first appearances, the sequence of organisms from bacteria and paramecia to porpoises and human beings is not grounded in phylogeny but in classification. The elision from phylogeny to classification is encouraged by the ease with which the branching structure of phylogenetic trees can be confused with taxonomic hierarchies. In phylogeny, species split successively through time; in classifications, kingdoms are divided into phyla, phyla into classes, classes into orders, until the species level is reached. What could be easier than setting up a simple isomorphic relation between these two branching structures? As theoretical taxonomists from Mayr (1942, 1969), Hennig (1959, 1966), and Simpson (1961b) to Eldredge and Cracraft (1980), Nelson and Platnick (1981), and Wiley (1981) have shown, the relationship between phylogeny and the taxonomic hierarchy is not as straightforward as the superficial similarity in structure between the two might imply. Perhaps phylogenetic development through time has been in some sense progressive, but this temporal dimension is systematically obscured in hierarchical classifications in which forms that have been extinct for hundreds of millions of years are classified alongside extant taxa. In phylogeny extant species appear arrayed along the top while extinct species occur at branching points back into the distant past. In classifications, all species, whether extinct or extant, appear equally as terminal taxa.

Darwin's doubts about the superiority of molluscs to crustaceans can be reiterated with respect to any sequence of taxonomic groups. In what phylogenetic sense is a sea anemone more progressive than an oak tree? What justification can one possibly have for claiming that a tiny planarian is more advanced than a starfish, or an amphioxus vastly superior to a praying mantis? Such sequences are generated by selecting groups here and there in the phylogenetic tree, regardless of their relative ages or phylogenetic connections, and ordering them in a series by means of a particular criterion. The trouble is that such series can be generated for *any* characteristic just so long as it is a convergence and not an evolutionary homology. For short periods of time, certain evolutionary homologies do increase, but such increases periodically regress and never last for long. Only when characters are treated as convergences, characters that can evolve time and again, can lengthy series be developed. All of the characteristics that are currently listed as candidates for progressive development in biological evolution have evolved numerous times. Perhaps organisms through the eons exhibit a net increase in complexity, but multicellularity and subsequent increases in complexity occurred repeatedly.

Directionality and Higher Taxa

Darwin interspersed his discussion of the progressive character of the evolutionary process with the warning that he could see no way to test such hypotheses. One feature of present-day discussions of such topics by paleontologists and evolutionary biologists is that they are trying to test their impressions about phylogeny and connect their findings with current views about the evolutionary process. Because "progress" is so obviously value-laden, present-day biologists have replaced it with "direction." It is difficult enough to decide whether biological evolution has a direction without attempting to decide whether or not any putative direction counts as being progressive. However, the relations between "progress" and "direction" are not all that straightforward. Direction obviously does not imply progress. For example, species diversity might well have increased through the millennia, suffering only an occasional setback. Hence, one might legitimately claim that biological evolution has had a direction without being committed to the view that increases in species diversity are in any value-laden sense "better."

One might think that progress entails direction. Without evolution having at least a direction, how can it possibly be progressive? However, according to a long list of criteria, human beings "reluctantly" concede that they are the highest form of life on Earth. With these criteria, evolution is progressive, but no direction can be discerned on several of them prior to the appearance of *Homo sapiens*. For example, throughout the vast reaches of geological time, no indication exists of any other organism being capable of language use in the sense applicable to Man. Then, at the last minute, Man evolved, and with him "true" language. As far as net direction is concerned, this sequence exhibits an increase, indicating exactly how weak the notion of net direction is. As anthropocentric and vacuous as the conclusion might be, we nevertheless find the emergence of language "genuine progress." With respect to the characteristics unique to man, evolution exhibited neither direction nor progress throughout most of its history. It became progressive only quite recently with the evolution of its crowning achievement. This is progress made easy.

Although the notion of direction in biological evolution may not be value-laden, it is theory-laden. Once again, two questions are at issue: does the fossil record reveal directionality of any sort, and do current views about the evolutionary process imply that it should? One might expect the first question to be easier to answer than the second. After all, the presence of certain patterns in the fossil record is a matter of empirical fact. Either the numbers of marine taxa have remained roughly constant since the Cambrian, or they have not. This view of the matter turns on a belief that patterns of empirical phenomena can be discerned in the absence of any theories about the processes that produced these patterns, a view that is currently anathema to present-day philosophers of science, this one included (Hull 1986; see also Gould 1987 and this volume).

Throughout the history of science, philosophers of science as well as scientists themselves have periodically claimed that scientists can and should stick to the facts and nothing but the facts. Static patterns in empirical phenomena can be observed; natural processes cannot. Processes must be inferred, and the only basis for such inferences are observed patterns. Only after a period of extensive fact gathering are scientists entitled to speculate about the processes that produced the patterns they have discerned. Allowing one's speculations about process to influence the very structuring of the data that will be used to test these speculations is, so we are told, viciously circular.

As appealing as this line of reasoning has been in the past and continues to be to the present, it is seriously flawed. General arguments can be brought to bear on such inductivist philosophies of science, but the observation that counts against it most decisively is that this way of proceeding in science has been singularly unsuccessful. Those scientists who have attempted to produce theory-free descriptions of natural phenomena have had pitifully little impact on the course of science. As illogical as the method of reciprocal illumination may be, it has been the method that the most productive scientists throughout the history of science have utilized. The feedback between pattern and process is messy; it affords ample opportunity for errors, not to mention the amplification of these errors, but it is the only method to date that has led to sustained success.

If the recognition of taxonomic groups were a theory-free activity, then biologists could simply count taxa to see which patterns emerge. Even though counting taxa under such a supposition would still confront numerous technical difficulties, one complication would be avoided: the feedback between observation and theory. However, this complication cannot be avoided. In spite of generations of debate over taxonomic philosophy, the vast number of practicing systematists do not use any explicitly stated, formal set of principles and methods in the recognition and ordering of taxa. They learn their craft from an experienced systematist in the sort of apprentice system that has characterized science from its beginning. As Simpson (1961b) noted long ago, biological classification throughout its long history has been as much an art as a science. Some systematists take relative numbers of taxa into account, raising large taxa to higher category levels; others do not weight such considerations very heavily. Some emphasize morphological gaps between taxa even if considerable diversity exists within the resulting taxa; others do not. The list of considerations that influence biological classifications is not indefinite, but it is long. As a result, anyone who simply counts the taxa to be found in the literature is liable to come up with very noisy data sets. The only patterns that are liable to be discernible are those that are extremely marked. The Permian mass extinction might well show up no matter how one counts, but less extreme events are liable to get lost in the static.

Although the effort would be superhuman, one would think that the ratio of noise to message in the fossil record could be reduced by reclassifying all organisms according to a single, consistent set of principles. The question then becomes, which set? Currently three different schools of systematics are prominent: pheneticists, cladists, and evolutionary systematists. According to pheneticists, taxa should be produced by clustering organisms by explicitly formulated quantitative techniques, using sufficiently large numbers of characters. The characters need not be putative evolutionary homologies, the resulting taxa need not have any relation to phylogeny, and the boundaries between taxa need not be sharp. In-principle arguments against phenetic taxonomy to one side, the major problem with phenetic methods is that they can be used to generate indefinitely many classifications, and such formal considerations as stability and robustness alone have not proven sufficiently powerful to pick out a single classification as preferable. Phenetic methods are extremely useful in uncovering a variety of patterns in one's data, but they provide insufficient means for deciding among these patterns. If taxa are to be counted, sooner or later a systematist must settle on a particular classification, like it or not.

The main strength of the other two schools of systematics is that they do present reasons in addition to purely formal considerations for preferring one classification to another. Cladistics in its purest form is built on a single empirical assumption: descent with modification. To the extent that phylogeny is largely a matter of successive splittings, it would seem that cladistic classifications should be hierarchical. A taxonomic boundary can be marked as surely by a single character as by a dozen, just so long as the characters used nest perfectly. Any departures from perfect nesting imply errors in character recognition. According to pattern cladists, as they have come to be known, pattern recognition among organisms need not and should not assume anything about the evolutionary process itself. It makes no difference whether natural selection is a major or minor directive force in evolution, whether meiosis is costly or not, and so on: the principles of cladistic analysis remain untouched. As in the case of pheneticists before them, even pattern cladists allow some considerations of process to influence their recognition of species, e.g., males and females must be included in the same species regardless of phenetic similarity (Sokal and Crovello 1970) or defining characteristics (Nelson and Platnick 1981). Mating and the production of offspring are processes; they are also processes that play important roles in evolution, but one can admit their a priori relevance to classification while excluding use of other broader evolutionary processes. However, some cladists join with evolutionary systematists in acknowledging important inferential relations between the patterns that they recognize in their classifications and the evolutionary process at large (e.g., Wiley 1981).

The major difference between evolutionary systematists and cladists of all sorts turns on the notion of monophyly. Pheneticists do not care whether their Operational Taxonomic Units are monophyletic or not. Both cladists and evolutionary systematists insist that their higher taxa be monophyletic. They simply mean different things by "monophyletic." Cladists insist that all species in a higher taxon be descended from a single ancestral species, and that all species descended from a single ancestral species be included in the same higher taxon. Ignoring the first criterion produces polyphyletic taxa; ignoring the second produces taxa that are paraphyletic. Evolutionary systematists reject the second criterion. They count paraphyletic taxa as being monophyletic. Some evolutionary systematists do accept the first criterion. They agree with cladists that all species included in a higher taxon must be traceable to a single ancestral species. Other evolutionary systematists reject even this much. Higher taxa can count as being monophyletic even though they are descended from more than one ancestral species just so long as these ancestral species are included in a taxon classified at its own or lower category level (Simpson 1961b).[2]

One justification for preferring clades is that the processes that produce them do produce them. Lineages do split successively through time (although mergers also occur, especially among plants). Even though phylogenies may be difficult to reconstruct, they do exist. So this justification goes, the patterns produced by phylogenetic development are primary because the processes that produce them are fundamental to our understanding of the living world. Just as the periodic table of elements is fundamental in classical physics, phylogenetic classifications are basic in biology. Evolutionary systematists agree with the preceding justification as far as it goes, but they insist that there is more to evolution than genealogy. They argue that the ecological dimension to the evolutionary process should also influence how we classify organisms. A purely cladistic classification is not good enough. Grades are just as real as clades. Sometimes one or more lineages invade a new adaptive zone and proliferate. Such grades as birds and reptiles should be interspersed among more narrowly defined clades (Mayr 1981).

Several important difficulties confront all cladists and all evolutionary systematists. At one end of the spectrum are pattern cladists who, like the pheneticists, strive to make classification as theory-neutral as possible. For them, the interplay between observation and theory is minimal. At the other end of the spectrum are evolutionary systematists who propose to let both genealogical and ecological considerations influence their classifications. They are put in the position of specifying precisely how they propose to accomplish these ends. Thus far, evolutionary systematists have not been all that clear about the ways in which grades are to be distinguished and how grades and clades are to be unambiguously integrated into a single classification.

However, for the purposes of this paper, the point I wish to emphasize is that the classifications produced by using these various taxonomic principles are sure to be quite different. Cladistic classifications differ drastically from both casual, intuitive classifications and from more self-conscious evolutionary classifications. Anyone who doubts this statement need only compare McKenna's (1975) classification of mammals with earlier classifications. Those biologists who are attempting to discern patterns in the fossil record are liable to come up with very different patterns, depending upon whether they are counting clades, grades, or some combination of the two. Any pattern in the fossil record that could survive such a reworking of the taxa being counted would be robust indeed.

Paleontologists and evolutionary biologists have discerned a variety of patterns in the fossil record, on the basis of the taxa recognized by hundreds of workers, using a heterogeneous assortment of largely unspecified taxonomic methods. Reworking these taxa according to a single, consistent methodology should produce a clearer picture of what went on in the past. But which "directions" are likely to survive this process? For example, Van Valen (1984) has discovered an exponential decrease in probability of extinction for families of marine organisms, interrupted only by the great Permian extinction. He concludes his paper, however, with the observation that a "cladistic classification would preclude any analysis like that of the present paper." Gould et al. (1987) have found another trend, one toward bottom-heavy clades, i.e., clades that exhibit more diversity during the first half of their existence than during the second half. These authors justify their use of clades rather than "heterogeneous totalities" by noting that evolution is fundamentally a genealogical process. However, they note that in actual practice they used "established taxonomic groups," which are anything but clades in the strict sense. To the extent that some of these families are paraphyletic, they will automatically be bottom heavy. As these authors note, a decision as to whether they have found a genuine temporal asymmetry in evolution or an artifact of paraphyletic classifications must await a fully cladistic reclassification of marine organisms.

Because this same conclusion holds for all apparent phylogenetic trends discerned among higher taxa, firm conclusions must await such consistent classifications. If the same direction emerges for monophyletic groups broadly defined, monophyletic groups narrowly defined, and heterogeneous totalities, then this is very likely an actual phylogenetic trend. However, if a trend emerges on the basis of one sort of classification but not the others, a decision must be made about which sort of classification to utilize. As circular as it may seem, one good reason for preferring a particular mode of classification is that regularities emerge when it is used. If these regularities are of the sort that one expects on the basis of our current understanding of the evolutionary process, then both this mode of classification and our current understanding of the evolutionary process gain mutual support. However, one possibility must not be ignored: if no trends in the fossil record can be discovered no matter how higher taxa are structured, then possibly evolution has no direction.

Directionality and Species

Traditionally paleontologists have limited themselves to discerning and naming higher taxa. Their basic units of analysis are orders, families, and genera, not species. However, if one plots the respective numbers of these higher taxa through time, different distributions result (see Sepkoski 1978, 1979, and Sepkoski et al. 1981). Perhaps these variations are due to differences in the way these higher taxa are constructed. One way around this problem is to narrow one's focus to species, the basic units of both biological classification and the evolutionary process. If species are the same across taxonomic groups as well as through time, then estimates of species richness should give an accurate reflection of any putative direction to evolution.

However, the recognition of species is not without its difficulties. Even in the best of circumstances, neontologists find the task of discerning genuine species quite time consuming. Sibling species and polytypic species are common. Although inferences from character distributions to reproductive gaps are not infallible, they can produce reasonably good first approximations -- if one has enough of the right sort of distributional data. Phenetic clusters are not good enough. If paleontologists had access to the same kinds of distributional data available to neontologists, then they could reason analogously, even though more direct means of testing reproductive gaps is denied them. Unfortunately, only in very special circumstances can paleontologists approach the same sort of warrant for their inferences to species status as that of neontologists.

Even so, some paleontologists think that they can identify species in the fossil record and can test earlier claims about the direction of biological evolution based on higher taxa. Signor (1985), for example, lists four models of species richness from the Cambrian to the present (see fig. 1): the Empirical Model (Valentine 1970), the Equilibrium Model proposed by Gould, Raup, Sepkoski, Schopf, and Simberloff (Gould et al. 1977), Bambach's (1977) species-richness Model, and the Consensus Model of Sepkoski, Bambach, Raup and Valentine (Sepkoski et al. 1981). Each of these models employs a different method of estimating species richness through time and generates a different curve. According to the Empirical Model, species richness increased a bit during the Devonian, decreased in the Triassic, and then took off at the end of the Mesozoic; in the Equilibrium Model, it increased rapidly after the Cambrian and leveled off thereafter. Bambach agrees that a leveling off in numbers of species occurred soon after the Cambrian, but his plateau is half as high as that proposed in the Equilibrium Model, and was followed in the Cenozoic by a rapid increase. Finally, representatives of each of the previous models combined forces to produce a Consensus Model. According to this model, species richness gradually increased after the Cambrian, decreased in the Jurassic, and increased thereafter.

If one plots data on higher taxa after the Cambrian, the distribution implied by the Equilibrium Model fits the relative number of orders reasonably well, the Consensus Model approximates the number of genera, while none of the models fits the distribution of families all that well. The Consensus Model comes closest (see fig. 1). But how about species? After doing his best to correct for such sources of error as sampling intensity, Signor (1985) comes up with an estimate of species richness that approximates most closely Valentine's Empirical Model. If Signor is right, then there has been no direction to evolution during the Phanerozoic with respect to species richness until quite recently. For 500 million years, not much happened, and then for the past 100 million years the number of species has increased dramatically. Of course,

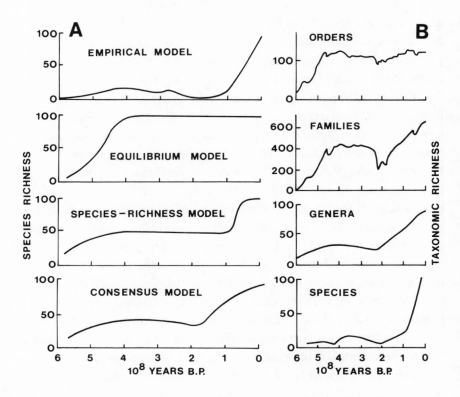

Figure 1. (a) Four models of species richness in the Phanerozoic. These models represent changes in standing species richness relative to the Recent. (b) Trends in estimated taxonomic richness through time. These diagrams are modified from Signor 1985.

with Signor's figures, there has been an overall net increase in species richness, but it does not take much to generate overall net increases of anything. For example, throughout the history of life on Earth, there has been a net increase in resistance to DDT, in the percentage of fluorocarbons in the earth's atmosphere, and in the rights of women. Although the time scale for these latter examples differs from that of species richness, the curves are the same. Considering such curves as having a direction seems more than a little misleading.

Even when we turn to hypotheses about the putative causes of the relative abundance of species and/or higher taxa through time, we do not get much help. According to the advocates of the Equilibrium Model, species packing is the chief cause for the shape of biological evolution. In the absence of any catastrophic influences such as Oort clouds, impacts with comets, etc., species richness should increase exponentially until the carrying capacity of the earth is reached. As more species evolve, more niches come into existence. Instead of widespread species with broad distributions populating the earth, more and more species with increasingly limited numbers and ranges are produced. However, according to the advocates of the Equilibrium Model, an upper limit must exist for species packing; exactly where this upper limit lies is a moot question. For marine invertebrates, it seems to have been reached soon after the Cambrian. Just as the Hardy-Weinberg law serves as the base line for evolution in population genetics, species packing sets the basic shape of evolution over paleontological time.

Departures from equilibrium must be explained by reference to various special circumstances. One of the most general of these special circumstances is provinciality due to tectonic forces. For example, a marked decrease in shallow seas and coastal marine habitats would surely cause a dramatic decrease in the number of species that require such environments. Another factor that might influence species richness is the sequential appearance of such evolutionary innovations as autotrophy, sexual reproduction, multicellularity, etc. Because most of these "evolutionary triggers" occurred prior to the Cambrian, they are not likely to greatly influence the Phanerozoic data (Schopf et al. 1973). The chief exceptions are successive invasions of land and air well after the Cambrian. However, as a simple inspection of figure 1 indicates, neither "descriptive" plotting of relative numbers of taxa nor prescriptive models about what the shape of biological evolution should be indicate much in the way of direction. Of course, they all show net increase.

Directionality and Organisms

Two problems remain with the above figures. First, living creatures appeared on Earth about four and a half billion years ago. The data discussed above involve only the last 650 million years. Initially, it would seem, biological evolution did not have much of a direction until the great Cambrian explosion. Prior to the Cambrian, the number of species was quite small. In fact, prior to the evolution of sexuality one might conclude, with some justification, that no species existed at all, and this is the second problem. A very common belief dating from Aristotle to the present is that every organism belongs to some species or other, and that the species category marks a level of organization characteristic of all organisms regardless of all else. When the first molecule capable of replicating itself appeared on Earth, the first species appeared. Granted that species status in sexual organisms is determined by reproductive gaps and that gene flow is extremely important in the evolutionary process, even so asexual organisms form species comparable to those in sexual organisms.

I find it difficult to reconcile the claim that sexual and asexual species are both species of the same sort with the repeated assertions about the overwhelming effect of sexual reproduction on the evolutionary process. If gene flow, recombination, increased genetic heterogeneity, etc., are so important in the evolutionary process, how can organisms that lack all these characteristics evolve in the same ways as those that possess them? Of all the levels of organization that have been recognized in the living world, why must only the species level be ubiquitous? Not all genes are organized into chromosomes, not all organisms are multicellular, not all organisms form colonies, etc. The notion that all organisms form species seems no more justified than claiming that all organisms are multicellular. Granted that a systematist must recognize taxonomic species, it does not follow that these classificatory necessities reflect entities that actually perform a function in the evolutionary process or even result from it.

One alternative is to conclude that species evolved when sexuality did. Mitosis evolved about two billion years ago, meiosis a half a billion years later. These innovations introduced new levels of organization. One consequence of this perspective is that estimates of species richness cannot be extrapolated into the distant past because no species existed for most of the history of life on Earth, only organisms. However, one might discover a direction to evolution in terms of number of organisms. Right now the number of sexual and asexual organisms are probably roughly the same. Even though must insects are sexual, most blue-green algae are not.[3] Of course, some gene exchange occurs in some blue-green algae, about one in 250 million replications. If that is all it takes for a group to count as "sexual," then sexuality is universal. But as in the case of net direction, such claims are all but vacuous. On this same line of reasoning, asexuality is equally universal. Even human beings occasionally produce identical twins. Since the production of two organisms from one without meiosis is the benchmark of asexual reproduction, then *Homo sapiens* should count as being asexual. I hope the readers of this paper can recognize a *reductio ad absurdum* argument when they see one.

The distinction between sexual and asexual organisms to one side, perhaps there has been an increase in the numbers of organisms and/or biomass for the past four or five billion years. Even though estimates of numbers of organisms or total biomass through time are, to say the least, impressionistic, Dobzhansky et al. (1977, p. 512) guess that on the average both have increased. However, putative directionality with respect to numbers of organisms and/or increased biomass is not nearly as interesting as the examples usually cited when evolution is claimed to have a direction.

The Need for Evolutionary Progress

When the person sitting next to me on a plane asks what I do, I never say that I am a philosopher because, if I do, I am immediately confronted with questions like, "What is the meaning of life?" As Provine (this volume) argues, there is no meaning to life, at least not one Big Meaning. There are, however, lots of little meanings. For most people, lots of meanings are not good enough. They want one meaning that will make life worthwhile for all people in all ages regardless of circumstance. Nothing else will do. In the absence of such a universal meaning, life is nothing but a black hole. The question of whether biological evolution has a direction is of this same sort. As the number of people who attended the symposium held at the Field Museum of Natural History on evolutionary ideas of progress in 1987 indicates, many people take this question to be of extreme importance: "Is biological evolution progressive? Does it have a direction?" The answer that biological evolution has not just a direction, but lots of them, is not likely to satisfy the need that gives rise to this question. I am afraid that this need cannot be fulfilled, either by the fossil record or by a theory of biological evolution.

Notes

1. Anyone who finds Gould's statement in this volume a bit extreme -- that progress is a "noxious, culturally embedded, untestable, nonoperational, intractable idea that must be replaced if we wish to understand the patterns of history" -- needs only to read Nisbet's Introduction and Epilogue. According to Nisbet (1980, p. 2), "Our problem in this final part of the twentieth century is compounded by the fact that the dogma of progress is today in the official philosophies or religions of those nations which are the most formidable threats to Western culture and its historical moral and spiritual values -- one more instance of the capacity for Western skills and values to be exported, corrupted, and then turned against the very West that gave them birth." The future of the idea of progress, as far as Nisbet (1980, p. 352) is concerned, depends on the future of Judaeo-Christianity in the West. "If the idea of progress does die in the West, so will a great deal that we have long cherished in this civilization" (Nisbet 1980, p. ix). Although Gould does not reject everything that has been willed to us by the past, he is likely to respond, "Good riddance to quite a bit of bad rubbish."

2. All sides of this dispute insist that they are using the term "monophyly" correctly and that their opponents are modifying long-standing usage. As important as such semantic contests are in science, this particular one is not relevant to the goals of this paper.

3. Although the sexual/asexual distinction coincides roughly with the distinction between prokaryotes and eukaryotes, the two are not the same.

References

Bambach, R. K. 1977. Species richness in marine habitats through the Phanerozoic. *Paleobiology* 3:152-67.

Bury, J. B. 1932. *The idea of progress*. New York: Macmillan.

Darwin, C. 1859. *The origin of species*. Cambridge: Harvard University Press.

Darwin, F., ed. 1903. *More letters of Charles Darwin*. London: Murray.

Dobzhansky, T., F. J. Ayala, G. L. Stebbins, and J. W. Valentine. 1977. *Evolution*. San Francisco: Freeman.

Eldredge, N., and J. Cracraft. 1980. *Phylogenetic patterns and the evolutionary process: Method and theory of comparative biology*. New York: Columbia University Press.

Glick, T. G., ed. 1974. *The comparative reception of Darwinism*. Austin: University of Texas Press.

Gould, S. J. 1977. Eternal metaphors of paleontology. In *Patterns of evolution as illustrated by the fossil record*, ed. A. Hallam, 1-26. New York: Elsevier.

Gould, S. J. 1987. Hatracks and theories. *Natural History* 90 (3):12-23.

Gould, S. J., N. L. Gilinsky, and R. Z. German. 1987. Asymmetry of lineages and the direction of evolutionary time. *Science* 236:1437-41.

Gould, S. J., D. M. Raup, J. J. Sepkoski, Jr., T. J. M. Schopf and D. S. Simberloff. 1977. The shape of evolution: A comparison of real and random clades. *Paleobiology*. 3:23-40.

Hennig, W. 1959. *Grundzuge einer theorie der phylogenetischen systematik*. Berlin: Deutscher Zentralverlag.

Hennig, W. 1966. *Phylogenetic systematics*. Urbana: University of Illinois Press.

Hull, D. L. 1967. The metaphysics of evolution. *British Journal for the History of Science* 3:309-37.

Hull, D. L. 1986. Les fondements epistemologiques de la classification biological. In *L'Ordre et la Diversite du Vivant*, ed. Pascal Tassy, 161-203. Fayard: Foundation Diderot.

Lovejoy, A. O. 1936. *The great chain of being*. Cambridge: Harvard University Press.

Mayr, E. 1942. *Systematics and the origin of species*. New York: Columbia University Press.

Mayr, E. 1969. *Principles of systematic zoology*. New York: McGraw-Hill.

Mayr, E. 1981. Biological classification: Toward a synthesis of opposing methodologies. *Science* 214:510-16.

McKenna, M. C. 1975. Toward a phylogenetic classification of the Mammalia. In *Phylogeny of the primates*, ed. W. P. Luckett and F. S. Szalay, 21-46. New York: Plenum Press.

Nelson, G., and N. Platnick. 1981. *Systematics and biogeography: Cladistics and vicariance*. New York: Columbia University Press.

Nisbet, R. 1980. *History of the idea of progress*. New York: Basic Books.

Schopf, J. W., B. N. Haugh, R. E. Molnar, and D. E. Satterwait. 1973. On the development of metaphytes and metazoans. *Journal of Paleontology* 47:1-9.

Sepkoski, J. J., Jr. 1978. A kinetic model of Phanerozoic taxonomic diversity. I. Analysis of marine orders. *Paleobiology* 4:223-51.

Sepkoski, J. J., Jr. 1979. A kinetic model of Phanerozoic taxonomic diversity. II. Early Phanerozoic families and multiple equilibria. *Paleobiology* 5:222-51.

Sepkoski, J. J., Jr., R. K. Bambach, D. M. Raup, and J. W. Valentine. 1981. Phanerozoic marine diversity and the fossil record. *Nature* 293:435-537.

Signor, P. W., III. 1985. Real and apparent trends in species richness through time. In *Phanerozoic Diversity Patterns*, ed. J. W. Valentine, 129-50. Princeton: Princeton University Press.

Simpson, G. G. 1961a. Lamarck, Darwin, and Butler: Three approaches to evolution. *American Scholar* 30:238-49.

Simpson, G. G. 1961b. *Principles of animal taxonomy*. New York: Columbia University Press.

Sokal, R. R., and T. Crovello. 1970. The biological species concept: A critical evaluation. *American Naturalist* 104:127-53.

Valentine, J. W. 1970. How many marine invertebrate fossil species? A new approximation. *Journal of Paleontology* 44:410-15.

Van Valen, L. 1984. A resetting of Phanerozoic community evolution. *Nature* 307:50-52.

Wagar, W. W. 1972. *Good tidings; The belief in progress from Darwin to Marcuse*. Bloomington: Indiana University Press.

Wallace, A. R. 1889. *Darwinism: An explanation of natural selection with some of its applications*. London: Macmillan.

Wiley, E. O. 1981. *Phylogenetics: The theory and practice of phylogenetic systematics*. New York: John Wiley.

Progress in Evolution and Meaning in Life

William B. Provine

"Progress might have been all right once, but it's gone on too long."
---Ogden Nash

If the title for this paper seems overly ambitious and pretentious, then consider the more accurate (but unwieldy) one: "Progress in evolution, the modern synthesis in evolution, the foundation of ethics, and meaning in life." Julian S. Huxley addressed all of these ponderous issues in the final chapter of *Evolution: The Modern Synthesis* (1942) that gave the name to the "Evolutionary Synthesis" of the 1930s and 1940s (Mayr and Provine 1980). Huxley argued that a grand synthesis had occurred in modern evolutionary biology, proving that the process of organic evolution in nature was purposeless but progressive. Humans were at the pinnacle of evolutionary progress; and although no purpose or direction from god was detectable in evolution, ethics could be based upon an understanding of evolutionary progress. Moreover, man's place in nature, at the pinnacle of the evolutionary process, gave deeper meaning to human existence than did anything else.

I will argue that Huxley's idea of progress in evolution is merely the imposition of his cultural values upon evolution, that the modern synthesis in evolution is scarcely a synthesis at all and should be renamed, that ethics cannot be founded upon any notion of "progress" in evolution, and that the process of evolution gives no meaning in life. Huxley was correct, however, that developments in evolutionary biology did destroy the widespread hope and belief that evolution was guided by conscious or unconscious purposive forces.

49

Julian Huxley and Progress in Evolution

Evolutionary biology raises fundamental issues about human culture. Almost simultaneously with his invention of the mechanism of natural selection, Charles Darwin opened his M and N notebooks on man, mind, and materialism (see especially the careful new critical edition of Darwin's notebooks in Barrett et al. 1987). In these he jotted down his musings on the cultural, psychological, and metaphysical implications of evolution by natural selection. Most of the major evolutionists of the nineteenth century wrote and lectured widely on the larger implications of evolution. The literature from the period 1859-1900 on the cultural implications of evolution is enormous. Julian Huxley's grandfather, Thomas Henry Huxley, was keenly interested in the implications of evolution. But he rejected the idea that ethics could be based upon the evolutionary process, which he considered to be totally amoral and frequently disgusting. He presented this argument forcefully in the second Romanes Lecture in 1893 ("Evolution and Ethics," T. H. Huxley 1894), when Julian was six.

From the beginning of his scientific career, Julian Huxley always and unwaveringly believed that evolution was progressive and offered hope and meaning to human existence. His first book, *The Individual in the Animal Kingdom* (1912), glorified from beginning to end the evolutionary progress of the individual, from the primordial protozoan individual to human society. Huxley even argued that "the State" was an evolutionarily advanced individual. He said that evolution tended to produce brains: "It is noteworthy that the course of internal differentiation has over and over again -- in worms, in insects, in crustacea, in spiders, in molluscs, and in vertebrates -- tended in the same direction -- towards the formation of a Brain" (Huxley 1912, p. 140). The book concluded with this note:

> All roads lead to Rome: and even animal individuality throws a ray on human problems. The ideals of active harmony and mutual aid as the best means to power and progress; the hope that springs from life's power of transforming the old or of casting it from her in favour of new; and the spur to effort in the knowledge that she does nothing lightly or without long struggle: these cannot but help to support and direct those men upon whom devolves the task of moulding and inspiring that unwieldiest individual -- formless and blind to-day, but huge with possibility -- the State. (Huxley 1912, p. 154)

I think this passage helps to clarify Huxley's otherwise mysterious defense of the writings of Teilhard de Chardin when they finally began to be published in the late 1950s. Teilhard's notion of progressive evolution that led to the "noosphere" and later to the collective consciousness of the Omega Point (Teilhard de Chardin 1959) was congenial to Huxley's life-long beliefs about the progressive development of individuals in evolution.

Judging from his praise of Bergson (Huxley 1912, p. vii) and of Orthogenesis (directed evolution) two years later (Huxley 1914, p. 560), I suspect that in the early 1910s, Huxley believed that evolution was not only progressive, but purposive. Later, he rejected any trace of purpose in evolution.

Evolution: The Modern Synthesis was an enormous compendium of research on evolutionary biology, mostly since 1900. The thrust of Huxley's argument in the book is easily seen in the first two paragraphs of chapter one:

> Evolution may lay claim to be considered the most central and the most important of the problems of biology. For an attack upon it we need facts and methods from every branch of the science -- ecology, genetics, paleontology, geographical distribution, embryology, systematics, comparative anatomy -- not to mention reinforcements from other disciplines such as geology, geography, and mathematics.
>
> Biology at the present time is embarking upon a phase of synthesis after a period in which new disciplines were taken up in turn and worked out in comparative isolation. Nowhere is this movement towards unification more likely to be valuable than in this many-sided topic of evolution; and already we are seeing the first-fruits in the re-animation of Darwinism. (Huxley 1942, p. 13)

Not until the last chapter, "Evolutionary Progress," do the larger implications of the modern synthesis in evolution begin clearly to emerge. Huxley defined evolutionary progress "as consisting in a raising of the upper level of biological efficiency, this being defined as increased control over and independence of the environment. As an alternative, we might define it as a raising of the upper level of all-round functional efficiency and of harmony of internal adjustment" (Huxley 1942, pp. 564-65).

Nothing about this definition of evolutionary progress indicated that evolution was purposive. Indeed, Huxley stated that "the ordinary man, or at least the ordinary poet, philosopher, and theologian" always was anxious to find purpose in the evolutionary process.

> I believe this reasoning to be wholly false. The purpose manifested in evolution, whether in adaptation, specialization, or biological progress, is only apparent purpose. It is just as much a product of blind forces as is the falling of a stone to earth or the ebb and flow of the tides. It is we who have read purpose into evolution, as earlier men projected will and emotion into inorganic phenomena like storm or earthquake. If we wish to work towards a purpose for the future of man, we must formulate that purpose ourselves. Purposes in life are made, not found. (Huxley 1942, p. 576)

Without purpose of some kind, however, how could evolution serve as a guide to ethics or give a sense of meaning in human life? Huxley answered:

> But if we cannot discover a purpose in evolution, we can discern a direction -- the line of evolutionary progress. And this past direction can serve as a guide in formulating our purpose for the future. Increase of control, increase of independence, increase of internal co-ordination; increase of knowledge, of means for co-ordinating knowledge, of elaborateness and intensity of feeling -- those are trends of the most general order. If we do not continue them in the future, we cannot hope that we are in the main line of evolutionary progress any more than could a sea-urchin or a tapeworm. (Huxley 1942, pp. 576-77)

By his own criteria, most evolutionary change was not progressive. Indeed, Huxley concluded that of all animals, only humans retained potential for further progress. All existing single-celled organisms were automatically ruled out from progressive evolution. "Only in the water have the molluscs achieved any great advance. The arthropods are not only hampered by their necessity for moulting; but their land representatives . . . are restricted by their tracheal respiration to very small size." Progressive evolution was impossible for cold-blooded animals; progress required both lungs and warm blood, "since only with a constant internal environment could the brain achieve stability and regularity for its finer functions." This left only the birds and mammals. "But birds were ruled out by their depriving themselves of potential hands in favour of actual wings, and perhaps also by the restriction of their size made necessary in the interests of flight." Huxley eliminated all mammals other than humans by arguments such as, "A horse or lion is armoured against progress by the very efficiency of its limbs and teeth and sense of smell: it is a limited piece of organic machinery" (all quotes in this paragraph from Huxley 1942, p. 570). Finally,

> The last step yet taken in evolutionary progress, and the only one to hold out the promise of unlimited (or indeed of any further) progress in the evolutionary future, is the degree of intelligence which involves true speech and conceptual thought: and it is found exclusively in man. . . . Conceptual thought is not merely found exclusively in man: it could not have been evolved on earth except in man. (Huxley 1942, pp. 570-71)

With this view of evolutionary progress, Huxley came to the logical conclusion: "Evolution is thus seen as a series of blind alleys" (p. 571), except, of course, for the evolutionary avenue leading to humans.

Working from the foundation of the new synthesis in evolutionary biology, Julian Huxley thus concluded, against the views of his grandfather, that organic evolution was progressive, and that evolutionary progress provided the basis for ethics and meaning in life.

The Evolutionary Synthesis

What exactly was this "modern synthesis" that figured so prominently in Huxley's book? As a historian I am immediately suspicious when anyone describes his or her views as the "new" or "modern" way of seeing things, to be sharply distinguished from the "old" inferior ways. The "new" billing is often little more than scholarly overstatement, an attempt to attract attention. Only two years earlier, Huxley had edited a volume that he entitled, *The New Systematics* (Huxley 1940). By all critical accounts then and now, this book had very little "new" systematics in it. So it is fair to ask, what was this "modern synthesis" in evolutionary biology?

About as many different versions of the evolutionary synthesis exist as there were major evolutionary biologists associated with it, augmented by a generous number of versions contributed by younger evolutionists, historians, and philosophers.

There is a wonderful symmetry in the interpretations of the evolutionary synthesis by the biologists who participated in it. Each felt his contribution to the synthesis was slighted, and that he had to fight for his proper place.

Ernst Mayr has said on many occasions that by the 1950s, the role of systematists and naturalists (his own work included) in the evolutionary synthesis had been terribly slighted by the geneticists, who argued that the central figures of the evolutionary synthesis were Fisher, Haldane, and Wright. Mayr began his campaign to focus attention on the contributions of the systematists with his Darwin centennial address, "Where Are We?", delivered at the Cold Spring Harbor Symposium (Mayr 1959). A major reason why he organized his symposium on the evolutionary synthesis in 1974 is precisely because he wanted to promote the role of systematists during the 1930s and 1940s. It may be difficult now to think that Mayr could have felt left out of the synthesis, but he did.

Sewall Wright wrote his strongly worded review of Mayr's "Where Are We?" address because he thought that Mayr was trying to read him and the other mathematical population geneticists out of the evolutionary synthesis (Wright 1960). Wright's feeling of being left out had already been exacerbated by the increasingly negative reactions to the importance of random genetic drift in evolution and the misunderstandings of his shifting balance theory of evolution (see Provine 1986, chap. 12). Wright therefore believed he was left out of the synthesis by both geneticists and systematists. Much of the disagreement between Mayr and Wright concerns not so much differences in their views of evolutionary biology, but their differences in interpreting each other's role in the evolutionary synthesis.

Julian Huxley sought to gain his place in the evolutionary synthesis by writing *Evolution: The Modern Synthesis* and by defending this work on every possible occasion. He never revised the book, but reissued it as a new edition twice more with new introductions. In the introduction to the 1962 (second) edition, he clearly wanted the world to know that he was the real architect of the new synthesis (see also his address to the Golden Jubilee meeting of the Genetics Society of America in 1950). He never thought he was given proper credit for his role in the synthesis.

George Gaylord Simpson hesitated to come to Mayr's conference on the evolutionary synthesis, suggesting that the conference setting was not conducive to elucidation of his role in the synthesis. After the conference, he was offended by the lack of appreciation of his contributions to the evolutionary synthesis, which he outlined more carefully in his autobiography, *Concession to the Improbable* (1978) and in his new introduction to the reprint of his *Tempo and Mode in Evolution* (1984). Simpson clearly believed that he was one of the truly major figures of the evolutionary synthesis, but that his role and that of paleontology had been slighted.

C. H. Waddington felt so slighted and left out of a prominent role in the evolutionary synthesis by Mayr and others that he published an entire book, *Evolution of an Evolutionist* (1975), demonstrating the importance of his role in bringing embryology into the synthesis. G. Ledyard Stebbins has remarked on many occasions that the role of botany in the evolutionary synthesis has been overshadowed by zoological contributions, partly from the forcefulness of zoologists in taking their share of the credit, aided by the deep-seated animal chauvinism so ever-present in our culture. Among the other major figures of the evolutionary synthesis, Dobzhansky, Fisher, Ford, Goldschmidt, Darlington, Muller, Rensch, and Timofeef-Ressovsky all believed their work in the evolutionary synthesis was not, in hindsight, given the full deserved credit. The symmetry here is ever so reminiscent of an academic department of top-notch scholars, each of whom predictably believes that his or her work is not properly appreciated by the chairman and the rest of the department, no matter what rewards, accolades, and support are provided.

Among younger evolutionists, disagreement about the evolutionary synthesis is great. Niles Eldredge has written that the evolutionary synthesis did occur, but did not go nearly far enough. He argues that the hierarchical taxa at and above the level of species were not properly synthesized with other evolutionary processes in the 1930s and 1940s (Eldredge 1985). Stephen Jay Gould argues that the evolutionary synthesis hardened, in its later stages, into a panselectionist outlook that minimized the multiplicity of evolutionary processes, including biological constraints, nonadaptive mechanisms, extremely rapid speciation, and species level selection (Gould 1983). Motoo Kimura says that the evolutionary synthesis occurred and is the reason for the intensely negative reaction to his neutral theory of molecular evolution from many of the living architects of the synthesis (Kimura 1983). Stebbins and Ayala (1981) argued strongly that the evolutionary synthesis occurred and provides a robust basis for evolutionary biology today, a view also defended by Futuyma (in press).

Janis Antonovics, in his presidential address to the American Society of Naturalists in 1986, argued that the evolutionary synthesis did occur but hindered rather than promoted the advance of evolutionary biology.

> My thesis is that the Evolutionary Synthesis failed in many serious and insidious ways. I propose that the Synthesis had little direct impact on the progress of evolutionary biology as a discipline and that, at the conceptual level, it may even have hindered rather than furthered our understanding of evolution. Many of the negative effects of the Synthesis have lasted to this day in terms of the institutional and conceptual structure of the field. I suggest that it is probably time that, rather than trying to finish the Synthesis as Eldredge (1985) has exhorted, we instead earnestly work to dismantle it. I suggest that only by achieving a Dys-Synthesis can we free ourselves of many of the methodological, conceptual, and even socioreligious difficulties that plague evolutionary biology. (Antonovics 1987, p. 321)

And finally, Mayr has just reevaluated the evolutionary synthesis in an essay entitled, "On the Evolutionary Synthesis and After" (Mayr, in press). Here he argues that the recent critics of the evolutionary synthesis did not appreciate the firm foundation provided by the evolutionary synthesis, yet he also refers to the synthesis alternatively as a unification and as a consensus. Shortly, I will argue that this terminology is crucial.

One clear fact shines amid all this diversity of opinion about the evolutionary synthesis: all agree that *something important happened in evolutionary biology during the 1930s and 1940s*. Whatever it was had not happened by 1930, but had happened by the Darwin centennial in 1959. What exactly had happened? Can we characterize the "evolutionary synthesis" more precisely? I think we can.

Perhaps the easiest way to begin is by specifying what the evolutionary synthesis is not. First, it is scarcely a synthesis at all. According to Huxley and later to Mayr, many fields were part of the synthesis. Among these were ecology, genetics, paleontology, geographical distribution, embryology, systematics, comparative anatomy, and some mathematics. I can agree that there was a quantitative synthesis of Mendelian heredity and various factors that can change gene frequencies in populations. This was accomplished in the models of Fisher, Haldane, Wright, Hogben, Chetverikov, and others. Yet they all disagreed, often intensely, with each other about actual processes of evolution in nature, even when their models were mathematically equivalent.

Beyond this genuine synthesis, the rest of the "evolutionary synthesis" was mostly exercises in removing barriers, consistency arguments (these two are closely related) and forging a consensus. A lot of what mathematicians call "hand waving" was also involved, especially at the time Huxley coined the name, "the modern synthesis."

I realize that Mayr and Shapere (in Mayr and Provine 1980) and others have been willing to characterize consistency arguments and removal of barriers between fields as "synthesis," but I prefer to call these developments what they are: consistency arguments and removal of barriers, and to reserve for "synthesis" that which is actually synthesized. Of course consistency arguments and removal of barriers can be precursors, even necessary ones, before synthesis takes place.

Secondly, the synthesis is not characterized by startling or extraordinary new discoveries, concepts, or theories. Some candidates might be Fisher's "fundamental theorem of natural selection," Wright's shifting balance theory of evolution in nature and his surface of selective values, Mayr's founder effect and genetic revolution in relation to geographic speciation, Waddington's concept of an epigenetic landscape, and Muller's "ratchet." While each of these played a significant role in the period of the evolutionary synthesis, none fits the bill as a concept or theory around which the evolutionary synthesis was built or centered, in the way, for example, that Darwin built his theory of evolution in nature around the mechanism of natural selection.

Finally, the evolutionary synthesis of the 1930s and 1940s was not characterized by agreement on the mechanisms of evolution in nature. Although Fisher and Wright reached almost entire agreement on the mathematical consequences of their different quantitative models of the evolutionary process, they disagreed intensely about the relative weights of different variables in the evolution in nature (see Provine 1985). They could not agree on the relative roles of selection and random genetic drift, or on the population structure of natural populations. Fisher and Ford had little use for Mayr's concepts of founder effect and genetic revolution in speciation (Ford 1964). Late in the 1940s, as Gould has pointed out, there was a "hardening of the synthesis" toward a process dominated by deterministic natural selection; and this domination lasted until the mid- to late 1960s, when it began seriously to erode (Gould 1983). But this hardening came only at the

very end of the synthesis period, years after Huxley and Mayr thought
evolutionary biology was "synthesized." The evidence is overwhelming that
evolutionary biologists disagreed strongly about mechanisms of microevolution
during the synthesis period, and even more strongly about mechanisms of
speciation (Provine, in press).

If the evolutionary synthesis was not primarily a synthesis, was not
characterized by important new discoveries or theories, did not generate
agreement among evolutionary biologists about mechanisms of microevolution or
speciation, and yet happened, then what was it? What did happen to
evolutionary biology during the 1930s and 1940s that convinced its participants
of palpable advance during this time?

A Different Interpretation: The Evolutionary Constriction

A revealing clue comes from the history of ideas about heredity in the
late nineteenth and early twentieth centuries. In 1894, the French biologist
Yves Delage, Professor at the University of Paris, submitted for publication an
enormous manuscript on ideas about heredity. It was completely up-to-date,
with numerous references to literature published in 1894 (his introduction to
the first edition was dated December, 1894). The book, entitled *L'Hérédité et
les Grandes Problèms de la Biologie Générale* (Delage 1895), was a huge
compendium and analysis of all theories of heredity before 1894. Delage cited
and referred to Focke's *Die Pflanzen-Mischlinge* (1882), in which Mendel's
work was discussed, but Delage did not himself refer in any way to Mendel.
This was no surprise for 1895.

The book was well received and Delage prepared a second, "revised,
corrected, and augmented" edition published in 1903. The bibliography was
unchanged, but the book was studded with new (specially starred) footnotes
that referred to and discussed the literature from 1894 to 1901. Correns and
Tschermak are nowhere mentioned. There is a substantial discussion of de
Vries's theory of intracellular pangenesis and even some of his work up to
1900, but nothing about the rediscovery of Mendel. Bateson's *Materials for
the Study of Variation* (1894) was the most recent of his works discussed in
the book. Mendel himself would have been appalled, since Nägeli's theory of

the ideoplasm received fifty-one pages of analysis. Weismann's theory of heredity rated fifty-three pages of discussion. This book was a truly monumental achievement, poised on the edge before the rediscovery of Mendelian heredity.

Eight years after the second edition of Delage's book, three German textbooks on heredity appeared, by Erwin Baur (1911), Richard Goldschmidt (1911), and Valentin Haecker (1911). Baur did not mention Delage or his book, and the two others merely included a citation of the book in their bibliographies. By the third edition of his textbook in 1919, Goldschmidt had dropped all mention of Delage's book. None of the other textbooks on heredity published after 1905 (when the first edition of Punnett's *Mendelism* appeared) even so much as mentioned Delage or his book on heredity. Included in this list are the textbooks of Lock (1906), Bateson (1909), Walter (1913), and Castle (1916). If Mendelism had not arrived, my guess is that despite any existing national chauvinism, Delage's book would have been translated into English and German and would have been the single most cited book on heredity during the first two decades of the twentieth century. But Mendelism did arrive in 1900, and it overwhelmed the second edition of Delage's book even as it was published.

Mendelism shoved all the theories of heredity that came before 1900 into obscurity. Nägeli's theory, which deserved more than fifty pages of analysis in 1900, rated none after that. Instead, a theory that Nägeli rated as a combination of being overambitious, limited, and inconsequential became *the* theory of heredity. Ten years after its rediscovery, Mendelism so dominated the study of heredity that young persons scarcely ever heard about the earlier theories of heredity. De Vries's *Intracellular Pangenesis* received some attention, but mostly because of his role in the rediscovery of Mendelism and his mutation theory.

I think this example from the history of the study of heredity sheds some light on the situation in evolutionary biology during the same time. By the time of his death in 1882, Darwin had convinced biologists around the world that organic evolution had occurred. Louis Agassiz, who had retained his creationist beliefs, was already considered an anachronism at the time of his death in 1873. But if Darwin had made evolutionists of biologists, he was much less successful in convincing them that natural selection was the primary mechanism of evolution in nature.

By 1900, the variety of evolutionary mechanisms advocated by biologists was bewildering. Most evolutionists emphasized mechanisms other than natural selection (Bowler 1983). Nearly every theory of heredity that Delage had examined in his big book was associated with a theory of a mechanism of evolution. Primary among the reasons for the rejection of natural selection as the primary mechanism of evolution in nature was the absence of convincing examples and outright abhorrence of the mechanistic implications of a mechanism that was purposeless and opportunistic. By my own rough estimate, the majority of evolutionary biologists in the late nineteenth century believed in one or another purposive mechanism of evolution.

What happened to this wide array of theories of evolution when Mendelian heredity was rediscovered? Did theories of the mechanisms of evolution disappear along with the theories of heredity? Not only did they not disappear, new theories of evolution emerged to join those already existing, the most important being Hugo de Vries's mutation theory (1901-03). The array of theories of evolution available in 1907 can easily be seen by consulting Vernon L. Kellogg's masterful *Darwinism To-Day* (1907). And it can be compared with H. W. Conn's *Evolution To-Day* (1886) or his *Method of Evolution* (1900), to see that the first years of the century had not seriously depleted the ranks of theories of mechanisms of evolution. Conn's *Evolution To-Day* was even reissued unchanged in 1907. A republication in 1907 of a compendium on heredity first published in 1886 was almost unthinkable.

In 1909, Delage and a collaborator, Marie Goldsmith, published in French a review of theories of evolution. This book devoted one of twenty-two chapters to Mendelian inheritance and its significance for evolution. Since I was tempted to conclude that the obscurity of the second edition of Delage's book on heredity occurred in large part because of the chauvinism of English and American biologists, the fate of his book on evolution is instructive -- it appeared in English translation in 1912 with publishers in both London and New York. There was continuity of theories of the mechanisms of evolution through the turn of the century.

This plethora of evolution theories suffered almost the same fate of extinction as theories of heredity before 1900, but at a different time. Mendelism did not seriously affect purposive theories of evolution, and did not vanquish evolutionary theories based upon the inheritance of acquired characters. Instead, just as Mendelism routed the earlier theories of heredity, so the "evolutionary synthesis" routed earlier theories of the mechanism of evolution.

The evolutionary synthesis was not so much a synthesis as it was a vast cut-down of variables considered important in the evolutionary process. Beginning in the late 1910s, the theoretical population geneticists -- Fisher, Haldane, and Wright -- argued that evolution in nature could be modelled quantitatively (Provine 1978). They demonstrated clearly with their models that evolution within a population could be accounted for by quantitative relationships between a relatively few variables. The prestige of the mathematical models, the application of models by Dobzhansky and Wright to natural populations of *Drosophila pseudoobscura* and by Ford and Fisher to *Panaxia dominula* (on these developments see Provine 1986), and application of the new genetics to systematics by Huxley and Mayr and to paleontology by Simpson, all combined with the dying out of an older generation of evolutionists to yield what seemed like a new way of seeing evolutionary biology.

What was new in this conception of evolution was not the individual variables, most of which had long been recognized, but the idea that evolution depended on relatively so few of them. So has gone much of science. Phenomena that appear so complex as to baffle the imagination, whether the motion of the heart, action of the tides, or periodicity of the comets, have been shown by science to depend essentially upon surprisingly few variables. Instead of a synthesis, evolutionary biologists during the 1930s and 1940s came to agree resoundingly upon a relatively small set of variables as crucial for understanding the evolution in nature. This I will now call the "evolutionary constriction," which seems to me to be a more accurate description of what actually happened to evolutionary biology.

The term "evolutionary constriction" helps us to understand that evolutionists after 1930 might disagree intensely with each other about effective population size, population structure, random genetic drift, levels of heterozygosity, mutation rates, migration rates, etc., but all could agree that these variables were or could be important in evolution in nature, and that purposive forces played no role at all. So the agreement was on the set of variables, and the disagreement concerned differences in evaluating relative influences of the agreed-upon variables. I agree with Gould that evolutionary biology "hardened" toward a selectionist interpretation especially during the late 1940s and 1950s. I see this as a further constriction of the evolutionary constriction (but I like the sound of "hardening of the constriction").

The evolutionary constriction drove from evolutionary biology all of the purposive theories of evolution that had been so common and popular before 1930. After the constriction, evolutionary biology was utterly devoid of purposive mechanisms. Thus one effect of the constriction was to make the conflict between evolution and religion inescapable, or put another way, the previously respectable compatibility of religion and evolution became less tenable. At the 1925 Scopes "Monkey Trial" in Dayton, Tennessee, the dean of American evolutionists was Henry Fairfield Osborn of the American Museum of Natural History. For the trial, he wrote a little book entitled, *The Earth Speaks to Bryan* (Osborn 1925), in which he argued that no conflict existed between evolution in nature and his own deep-seated Christian beliefs.

> If Mr. Bryan, with an open heart and mind, would drop all his books and all the disputations among the doctors and study first-hand the simple archives of Nature, all his doubts would disappear; he would not lose his religion; *he would become an evolutionist.* (Osborn 1925, pp. 20-21)

Osborn's own theory of the primary mechanism of evolution was a purposive force he called "aristogenesis." The evolutionary constriction eliminated this force along with Bergson's enormously popular *élan vital* and a host of other purposive theories. The argument from design, which had survived in evolutionary biology as long as Darwin's natural selection was supplemented by additional purposive mechanisms, withered after the constriction.

Julian Huxley, the Evolutionary Constriction, and Progress

Julian Huxley's *Evolution: The Modern Synthesis* is the perfect exemplification of my thesis concerning the evolutionary constriction. His book is anything but a synthesis. It consists almost entirely of a compilation and discussion of the constricted set of variables, with very little digestion or synthesis of the variables. (To be sure, Huxley's set of variables was larger than that employed by Dobzhansky, Wright, Simpson, Rensch, or any of the major figures of the period). Aside possibly from Huxley's endorsement of the extra-sensory perception experiments of J. B. Rhine (Huxley 1942, p. 574), there is absolutely no hint of purpose in any of the mechanisms of evolution presented in the book.

One of the great attractions of purposive theories of evolution before the constriction was the possibility of having a robust notion of evolutionary progress. The difficult trick was to have the progress without the purpose. That is what Huxley desperately wanted. His solution was to specify anthropomorphic criteria to measure progress, or in other words to argue that whatever leads to humans in evolution constitutes progress. Huxley vociferously denied any hint of anthropomorphism, yet was blatantly anthropomorphic in his discussion of evolutionary progress.

What represents progress in evolution? Increasing brain size with respect to body weight, longevity of taxa, degree of adaptive fit with environment, reservoir of genetic variability, ability to adapt quickly to environmental change, degree of sociality, and any number of other criteria come to mind. The family of AIDS viruses fits some of these criteria better than hominids. Judging from particular criteria, many evolutionary changes may be rationally and coherently interpreted as progressive or regressive. The problem is that there is no ultimate basis in the evolutionary process from which to judge true progress. When there was purpose in evolution, there *was* such a basis, but the evolutionary constriction eliminated the purpose.

When Thomas Henry Huxley argued that evolution in nature provided no basis for ethics, he fell back upon the Judaeo-Christian heritage. Julian Huxley gave up the Judaeo-Christian heritage very early. He seemed to think that the only hope for a foundation for ethics and deep meaning in life was tied somehow to progress in evolution. Thus giving up progress in evolution was for him giving up a great deal.

With this perspective, Huxley's laudatory introduction to Teilhard de Chardin's posthumous work, *The Phenomenon of Man* (1959), is understandable. Teilhard's purposive view of evolution seemed ridiculous to most evolutionary biologists on the centennial of the publication of Darwin's *On the Origin of Species*. Simpson (1964), Medawar (1961), and many others lambasted the book. Yet Julian Huxley was defending it, even after having argued for decades that

no purposive forces exist in evolution. The key was progress. Teilhard
believed there was progress in evolution and that it gave meaning to life. As
Huxley declared in the first paragraph of his introduction to the book,
Teilhard

> ... is able to envisage the whole of knowable reality not as a static
> mechanism but as a process. In consequence, he is driven to search for
> human significance in relation to the trends of that enduring and
> comprehensive process; the measure of his stature is that he so largely
> succeeded in the search. (Teilhard de Chardin 1959, p. 11)

Dobzhansky also defended Teilhard (Dobzhansky 1969), something that surprised
those who were unaware that he had always been a religious person who
wanted to find some ultimate meaning of life in evolution.

My assessment is that Huxley (and Dobzhansky) knew there was no
purpose in evolution but nevertheless clung to the forlorn hope that even
purposeless evolution might yield the same congenial meaning to life offered by
purposeful evolution. The vast majority of evolutionary biologists now agree
with the assessments of Simpson and Medawar, and look upon Huxley's
endorsement of Teilhard as an aberration and wholly out of step with modern
evolutionary biology.

When there was purpose in evolution, there could be real progress. The
evolutionary constriction ended all rational hope of purpose in evolution. That
is why I placed the quote from Ogden Nash at the beginning of this paper. It
belongs here: "Progress might have been all right once, but it's gone on too
long."

Implications of the Evolutionary Constriction

What modern evolutionary biology tells us is very different from what
Osborn believed the study of nature would tell Bryan. It tells us (and I would
argue that the same message flows in from physics, chemistry, molecular
biology, astrophysics and indeed from all of modern science) that nature has
no detectable purposive forces of any kind. Everything proceeds purely by
deterministic or stochastic processes, or more accurately by their interactions.
Quantum mechanical indeterminacy may exist and be absolutely fundamental,

allowing no hope of advancing beyond a probabilistic description of nature at the quantum level. But what all this means is the ultimacy of chance, although it should always be remembered that chance processes at the quantum level become the determinism of statistical laws at the level of many quanta, where all biology rests. Science reveals to us only chance and necessity, as Jacques Monod (1971) argued.

Modern science directly implies that the world is organized strictly in accordance with deterministic principles or chance. There are no purposive principles whatsoever in nature. There are no gods and no designing forces that are rationally detectable. The frequently made assertion that modern biology and the assumptions of the Judaeo-Christian tradition are fully compatible is false.

Second, modern science directly implies that there are no inherent moral or ethical laws, no absolute guiding principles for human society.

Third, human beings are marvelously complex machines. The individual human becomes an ethical person by means of only two mechanisms: deterministic heredity interacting with deterministic environmental influences. That is all there is.

Fourth, we must conclude that when we die, we die and that is the end of us. Bertrand Russell put it very nicely. He said:

> I believe that when I die I shall rot, and nothing of my ego will survive. I am not young, and I love life. But I should scorn to shiver with terror at the thought of annihilation. Happiness is nonetheless true happiness because it must come to an end, nor do thought and love lose their value because they are not everlasting. . . . Even if the open windows of science at first make us shiver after the cozy indoor warmth of humanizing myths, in the end the fresh air brings vigor and the great spaces have a splendor of their own. (Russell 1957)

The last sentence of Russell's quote is very poetic, smacking of baseless hope, but the first part is right on target. When I die I shall rot and that is the end of me. There is no hope of life everlasting. Even card-catalog immortality (a big enough section of a library catalog to be noticed by future scholars) is ephemeral at best.

Finally, free will, as traditionally conceived, the freedom to make uncoerced and unpredictable choices among alternative possible courses of action, simply does not exist. That assertion is often more difficult to accept than the assertion that there is no god, and for good reason. Many ethical traditions have been built upon atheism, but most of them are founded upon some secular notion of humans having the free will to make moral choices and thus bear true moral responsibility. What modern science tells us, however, is that human beings are very complex machines. If the evolutionary constriction is taken seriously, the evolutionary process cannot produce a being that is truly free to make choices.

Ethical philosophers generally disagree strongly with my argument that free will does not exist. One group of them, the majority, argues that no contradiction whatever exists between determinism and free will. This group argues that free will is like the freedom of a weather vane to turn freely in the wind, or like free-wheeling in a car. This argument reminds me of the fishmonger who shouted for hours "fresh fish for sale, fresh fish for sale." Toward the end of the day, a customer, attracted by the nicely shouted advertisement, discovered that the fish for sale smelled bad, and told the fishmonger that the fish were not fresh as he was shouting. The fishmonger replied, "Fresh fish is just their name. I never said the fish were fresh." Those who shout the compatibility of determinism and free will are merely calling human decision-making "free will." The direction a weather vane points is determined by the direction of the wind and is in no way free. Free-wheeling in a car simply means that the drive train is disconnected from the differential and the speed of the car is determined by its mass, air resistance, resistance in the mechanisms of the car, and the force of gravity on whatever incline the car happens to be.

Other ethical philosophers, who actually believe in free will, argue for an incompatibilist position on determinism and free will. These philosophers think that nature exhibits both determinism and indeterminism, as in quantum mechanical indeterminism. They argue with good reason that determinism and free will are incompatible. But they also believe in free will. This is a very difficult position to take because none of them can figure out a direct way that any of the indeterminism that we know about in nature could possibly, much less plausibly, produce human free will. So they use a backhanded logical argument that proceeds like this: If we had no free will, then there could be no moral responsibility. But some people are morally responsible. Therefore, some persons have free will, which, since it cannot come from determinism, must emanate from the indeterminism in nature.

This argument is appealing but demonstrably weak. The weak assumption is that without free will, no moral responsibility could exist. The only way that biological organisms can exhibit moral responsibility is for them to be trained to have it. We humans spend a vast amount of time trying (not very efficiently, I think) to make our children into morally responsible individuals. Moral responsibility is wholly compatible with no free will, indeed, is impossible with it.

My conclusion about these two warring schools of ethical philosophers is that they are both right. Those who are compatibilists about free will and determinism are right that the other group cannot get free will from indeterminism in nature. The second group is correct that the first cannot get free will from determinism. The reasonable conclusion, the one that is consonant with modern science, is that humans choose, but not freely. This view is advocated by only a tiny minority of brave ethical philosophers, all of whom deserve our sympathy for enduring the barbs of their colleagues.

Attempts to Escape the Inevitable

The squirming and squiggling to escape the implications of modern science are something to behold, and just as interesting among scientists as among federal judges, humanists, or liberal theologians who distrust science. I will begin with the United States National Academy of Sciences, one of the most prestigious groups of scientists in the world. A recent booklet on science and creationism published by the Academy opened with a preface signed by the president, Frank Press. He stated: "It is false, however, to think that the theory of evolution represents an irreconcilable conflict between science and religion. A great many religious leaders and scientists accept evolution on scientific grounds without relinquishing their belief in religious principles" (Press 1984). There is no conflict between science and religion according to the National Academy of Sciences.

At the recent Arkansas creationism trial (*McLean v. Arkansas Board of Education*, 1982), Judge William R. Overton said in his decision that we must follow the version of the Establishment Clause of the first amendment to the

Constitution found in *Lemon v. Kurtzman* (1973), which stated a three part requirement. First, a statute must have a secular legislative purpose. Second, the principal or primary effect of the statute must be one that neither advances nor inhibits religion. And third, the statute must not foster and excessive government entanglement with religion. Judge Overton ruled that the Arkansas statute mandating the teaching of creation "science" failed on all three counts (Overton 1982). His decision was applauded by scientists, theologians, religious leaders, sociologists, philosophers, and jurists.

At the trial and later, creationists argued that if the principal or primary effect of a statute must be one that neither advances nor inhibits religion, then it must be unconstitutional to teach evolution. Evolutionism contradicts our religion and thus inhibits our religion. Judge Overton's response at the trial was to argue that evolution was in no way antithetical to religion: "Evolution does not presuppose the absence of a creator or God. The plain inference conveyed [by the creationists] is erroneous." And he added a footnote:

> The idea that belief in a Creator and the scientific theory of evolution are mutually exclusive is a false premise and is offensive to the religious views of many. Dr. Francisco Ayala, a geneticist of considerable renown and a former Catholic priest who has the equivalent of a Ph.D. in theology, pointed out [to this court] that many working scientists who subscribed to the theory of evolution are devoutly religious. (Overton 1982, p. 943)

We have now seen the same argument twice, once from the National Academy of Sciences and once from a federal judge. The argument is that some evolutionary biologists are also religious, so there cannot be conflict between modern science and religion. Instances of the invocation of this argument are found everywhere. I have two objections to the argument. The first is that very few truly religious evolutionary biologists remain. Most are atheists, and many have been driven there by their understanding of the evolutionary process and other science. Second, it is not a compelling argument for the compatibility of science and religion that some few evolutionary biologists see no conflict. Most compatibilists I know are effective atheists, sometimes without realizing it.

According to the unlikely coalition of scientists, jurists, educators, theologians, and religious leaders, the conflict between science and religion exists only in the naive, literalist minds of the creationists. Scientists testify that creation science is stupid and so is the literalist religion of the creationists. The federal courts and even now the Supreme Court (in *Edwards v. Aguillard*, 1987) say the same things. So do the religious leaders. Nearly half of the American public is being called stupid or at best misguided. I agree that evolution has occurred and that creationists are wrong about that. But they are right that there is a conflict between science and religion, not only their religions.

The conflict is fundamental and goes much deeper than modern liberal theologians, religious leaders and scientists are willing to admit. Most contemporary scientists, the majority of them by far, are atheists or something very close to that. And among evolutionary biologists, I would challenge the reader to name the prominent scientists who are "devoutly religious." I am skeptical that one could get beyond the fingers of one hand. Indeed, I would be interested to learn of a single one. Osborn, Lack, Dobzhansky, and Fisher are dead, and no generation of compatibilists among prominent evolutionary biologists has replaced them.

I suspect there is a lot of intellectual dishonesty on this issue. Consider the following fantasy: the National Academy of Sciences publishes a position paper on science and religion stating that modern science leads directly to atheism. What would happen to its funding? To any federal funding of science? Every member of the Congress of the United States of America, even the two current members who are unaffiliated with any organized religion, professes to be deeply religious. I suspect that scientific leaders tread very warily on the issue of the religious implications of science for fear of jeopardizing the funding for scientific research. And I think that many scientists feel some sympathy with the need for moral education and recognize the role that religion plays in this endeavor. These rationalizations are politic but intellectually dishonest.

Many theologians have reacted to the rise of modern science by retreating from traditional conceptions of God and its presence in the world, calling this a more sophisticated view. God used to be all around us earlier in our cultural history. It used to perform miracles. It used to guide its people. People could detect God's presence all the time; but times have changed. God is more remote today. In fact, one cannot rationally discover anything that God does in the world anymore. A widespread theological view now exists saying that God started off the world, props it up and works through laws of nature, very subtly, so subtly that its action is undetectable. But that kind of God is effectively no different to my mind than atheism. To anyone who adopts this view I say, "Great, we're in the same camp; now where do we get our morals if the universe just goes grinding on as it does?" This kind of God does nothing outside of the laws of nature, gives us no immortality, no foundation for morals, or any of the things that we want from a God and from religion.

Meaning in Life

A friend of mine, John F. Haught, who teaches theology at Georgetown University in Washington, D.C., has written several books arguing for the compatibility of religion and science. He is very keen on having cosmic meaning in life (see especially Haught 1984). He has a logical argument that is reminiscent of the one I presented earlier about free will and moral responsibility, and it runs like this: Without cosmic meaning, our lives would be meaningless; but some humans have very meaningful lives; thus there must be cosmic meaning.

I can see no cosmic or ultimate meaning in human life. The universe cares nothing for us and will probably either continue to expand and cool, leading to an extermination of all living creatures in the universe, or the universe will cease to expand and will begin closure, which will result in everything crashing together in an unbelievably small space, thus obliterating all life. Even if the universe is in some sort of equilibrium and both scenaria above are wrong, the universe has so far exhibited no care for humans and gives no rational hope of future caring.

Humans are as nothing even in the evolutionary process on earth, and only a few individuals are remembered for as many as ten generations. There is no ultimate meaning for humans.

But certainly humans can lead meaningful lives. My own life is filled with meaning. I am married to an intellectually talented and beautiful woman, have two great sons, live on a 150 acre farm with pond, river, wild turkeys, and lots of old but good farm machinery; I teach at a fine university with excellent students and have many wonderful friends. But I will die and soon be forgotten. Jack Haught will have a tough time convincing me that my life is meaningless just because there is no cosmic meaning for it, or that the meaning that makes my life worthwhile is really cosmic meaning.

Foundation for Ethics

We need to recognize what modern science has done to us and try to come to terms with its implications for the foundation of morality. A growing number of young persons no longer believe in a god who lays down moral laws for them and punishes them if they disobey it. Many of them conclude that they can therefore behave in any way they can get away with. At the same time, we are surrounded by terrible moral behavior by adults. The only approach that I can hope will help the growing crisis in morality is through understanding that we humans are just complex machines without free will that have been poorly programmed for moral behavior. We are abysmally ignorant of the moral development of small children and are repulsed by the notion that they will grow up to be the moral persons we program them to be. B. F. Skinner has argued for many decades that we can really control the moral development of small children. In a general sense he is surely correct no matter how repulsive is the truth. We must learn more about how to make little children more surely into moral beings and help parents like me and my wife to better program our children. If this seems repellent, is it any better to do the same thing anyway, but do a much worse job of it because we refuse to think about what we are really doing?

We also must encourage people, especially young people, to think rationally about moral behavior. Their reason for behaving well should not be fear of a vindictive, purposive moral force that guards morality. They should behave well because if they do not, they are behaving foolishly in terms of their own interests. If you are unkind to others, then you reinforce other people's unkind behavior, and you are helping to generate a very poor society in which to live. It is in our own best interest to behave kindly toward

others. We can get more out of life with good and supportive friends.
Indeed, since there is no almighty friend, the only ones we have are biological
organisms or other machines (I would have said humans if I didn't know some
people who prefer dogs, cats, tractors, cars, birds, and even frogs over humans
as friends).

I grew up in Nashville, Tennessee, and frequently went to the Grand Ole
Opry. I remember particularly one gospel song with this line in the chorus:
"If you don't love your neighbor, then you don't love God." I think that what
modern science tells us is: "If you don't love your neighbor, then you're just
plain stupid!"

Conclusion

There is no ultimate progress in evolution. Evolution reveals no hint of
any purposive or guiding forces. Evolution provides no foundation for ethics
and no deep meaning in life. Julian Huxley's dream that evolution could
provide the basis for ethics and give meaning to life is just that -- a dream.
Nevertheless, the evolutionary process has produced in humans the possibility
for effective ethics and meaningful lives, though both are proximate rather
than ultimate.

References

Antonovics, J. 1987. The evolutionary dys-synthesis: Which bottles for which wine? *The American Naturalist* 129:321-31.

Barrett, P. H., P. J. Gautrey, S. Herbert, D. Kohn, and S. Smith, eds. 1987. *Charles Darwin's notebooks: 1836-1844*. Ithaca: Cornell University Press.

Bateson, W. 1894. *Materials for the study of variation*. London: Macmillan.

Bateson, W. 1909. *Mendel's principles of heredity*. Cambridge: Cambridge University Press.

Baur, E. 1911. *Einführung in die experimentelle Vererbungslehre*. Berlin: Borntraeger.

Bowler, P. J. 1983. *The eclipse of Darwinism: The anti-Darwinian theories in the decades around 1900*. Baltimore: Johns Hopkins University Press.

Castle, W. E. 1916. *Genetics and eugenics*. Cambridge: Harvard University Press.

Conn, H. W. 1886. *Evolution to-day*. New York: Putnam.

Conn, H. W. 1900. *The method of evolution*. New York: Putnam.

Darwin, C. R. 1859. *On the origin of species*. London: John Murray.

Delage, Y. 1895. *L'Hérédité et les Grandes Problèms de la Biologie Générale*. Paris: Librairie C. Reinwald. Second edition, 1903.

Delage, Y., and M. Goldsmith. 1912. *The theories of evolution*. London: Frank Palmer and New York: B. W. Huebsch.

De Vries, H. 1889. *Intracellular Pangenesis*. Jena: Gustav Fischer.

De Vries, H. 1901-03. *Die Mutationstheorie*. 2 volumes. Leipzig: Von Veit.

Dobzhansky, T. 1969. *The biology of ultimate concern*. New York: New American Library.

Eldredge, N. 1985. *Unfinished synthesis: Biological hierarchies and modern evolutionary thought*. New York: Oxford University Press.

Focke, W. O. 1882. *Die Pflanzen-Mischlinge. Ein Beitrag zur Biologie der Gewächse*.

Ford, E. B. 1964. *Ecological genetics*. London: Methuen.

Futuyma, D. J. In press. Presidential address, American Society of Naturalists. *American Naturalist*.

Goldschmidt, R. B. 1911. *Einführung in die Vererbungswissenschaft*. Leipzig: Engelmann.

Gould, S. J. 1983. The hardening of the modern synthesis. In *Dimensions of Darwinism*, ed. M. Grene, 71-93. New York: Cambridge University Press.

Haecker, V. 1911. *Allgemeine Vererbungslehre*. Braunschweig: Vieweg.

Haught, J. F. 1984. *The cosmic adventure: Science, religion and the quest for purpose*. New York: Paulist Press.

Huxley, J. S. 1912. *The individual in the animal kingdom*. Cambridge: Cambridge University Press.

Huxley, J. S. 1914. The courtship habits of the great crested grebe (*Podiceps cristatus*); with an addition to the theory of sexual selection. *Proceedings of the Zoological Society of London* 35:491-562.

Huxley, J. S., ed. 1940. *The new systematics*. Oxford: Oxford University Press.

Huxley, J. S. 1942. *Evolution: The modern synthesis*. London: Allen and Unwin. Second edition, 1962.

Huxley, J. S. 1950. Genetics, evolution, and human destiny. In *Genetics in the 20th century*, ed. L. C. Dunn, 591-621. New York: Macmillan.

Huxley, T. H. 1894. Evolution and ethics. In T.H. Huxley, *Evolution and ethics and other essays*, 46-116. London: Macmillan.

Kellogg, V. L. 1907. *Darwinism to-day*. New York: Holt.

Kimura, M. 1983. *The neutral theory of molecular evolution*. Cambridge: Cambridge University Press.

Lock, R. H. 1906. *Recent progress in the study of variation, heredity, and evolution*. London: John Murray.

Mayr, E. 1959. Where are we? *Cold Spring Harbor Symposia on Quantitative Biology* 24:1-14.

Mayr, E. In press. On the evolutionary synthesis and after. In *Toward a new philosophy of biology: Observations of an evolutionist*, by E. Mayr. Cambridge: Harvard University Press.

Mayr, E., and Provine, W. B., eds. 1980. *The evolutionary synthesis*. Cambridge: Harvard University Press.

Medawar, P. B. 1961. Review of *The phenomenon of man*. *Mind* 70:99-106.

Monod, J. 1971. *Chance and necessity*. New York: Knopf.

Osborn, H. F. 1925. *The earth speaks to Bryan*. New York: Scribner's.

Overton, W. R. 1982. Decision in McLean versus the Arkansas Board of Education. Reprinted in *Science* 215:934-43.

Press, F. 1984. Preface. In *Science and creationism: A view from the National Academy of Sciences*. Washington, D.C.: National Academy Press.

Provine, W. B. 1978. The role of mathematical population geneticists in the evolutionary synthesis of the 1930s and 1940s. *Studies of the History of Biology* 2:167-92.

Provine, W. B. 1985. The R. A. Fisher -- Sewall Wright controversy. *Oxford Surveys in Evolutionary Biology* 2:197-219.

Provine, W. B. 1986. *Sewall Wright and evolutionary biology*. Chicago: University of Chicago Press.

Provine, W. B. In press. Founder effects and genetic revolutions in microevolution and speciation: An historical perspective. In *Genetics, speciation, and the founder principle*, ed. L. V. Giddings, K. Kaneshiro, and W. W. Anderson. New York: Oxford University Press.

Punnett, R. C. 1905. *Mendelism*. Cambridge: Bowes and Bowes.

Russell, B. 1957. What I believe. In B. Russell, *Why I am not a Christian and other essays on related subjects*. London and New York.

Simpson, G. G. 1944. *Tempo and mode in evolution*. New York: Columbia University Press. Second edition, 1984; with a new introduction.

Simpson, G. G. 1964. *This view of life*. New York: Harcourt, Brace, and World.

Simpson, G. G. 1978. *Concession to the improbable: An unconventional autobiography*. New Haven: Yale University Press.

Stebbins, G. L., and F. J. Ayala. 1981. Is a new evolutionary synthesis necessary? *Science* 213:967-71.

Teilhard de Chardin, P. 1959. *The Phenomenon of man*. With an introduction by Sir Julian Huxley. New York: Harper and Brothers.

Waddington, C. H. 1975. *Evolution of an evolutionist*. Ithaca: Cornell University Press.

Walter, H. E. 1913. *Genetics: An introduction to the study of heredity*. New York: Macmillan.

Wright, S. 1960. Genetics and twentieth century Darwinism: A review and discussion. *American Journal of Human Genetics* 12:365-72.

Can "Progress" be Defined as a Biological Concept?

Francisco J. Ayala

The process of evolution appears to be obviously progressive. The earliest organisms were no more complex than some bacteria and blue-green algae. Three billion years later, their descendants include orchids, bees, peacocks and human beings, which appear to be more complex, advanced, or progressive than their primitive ancestors.

Upon reflection the issue becomes less obvious, because what do we mean when we say there has been progress in the evolutionary process? Some evolutionary lineages do not appear progressive at all: living bacteria are not very different from their ancestors of two or three billion years earlier. In addition, extinction can hardly be progressive, yet most evolutionary lineages have become extinct. Still more, organisms may be progressive with respect to others. For example, bacteria are able to synthesize all their own components and obtain the energy they need for living from inorganic compounds; human beings depend on other organisms.

Aristotle and other philosophers of classical Greece put forward the notion that living organisms can be classified in a hierarchy going from lower to higher forms. A similar view is implicit in the Bible. The creation of the world as described in the book of Genesis contains the explicit notion that some organisms are higher than others, and implies that living things can be arranged in a sequence from the lowest to the highest, which is man. The Bible's narrative of the creation reflects the common sense impression that earthworms are lower than fish and birds, and the latter lower than man. The idea of a "ladder of life" rising from amoeba to man is present, explicitly or implicitly, in all preevolutionary biology (Lovejoy 1936).

The theory of evolution adds the dimension of time and genetic continuity, or history, to the hierarchical classification of living things. The transition from bacteria to humans can now be seen as a natural, progressive development through time from simple to gradually more complex organisms. The expansion and diversification of life can also be judged as progress; some form of advance seems obvious in the transition from one or only a few kinds of living things to the several million species living today.

A Definition of Progress

The meaning of statements like "Progress has occurred in the evolutionary sequence leading from bacteria to humans" or "The evolution of organisms is progressive" is not immediately clear. Such expressions may simply mean that evolutionary sequences have a time direction, or even more simply that they are accompanied by change. The term "progress" may be clarified by comparing it with other related terms used in biological discourse. These terms are "change," "evolution," and "direction."

The term "change" means alteration, whether in the position, the state, or the nature of a thing. Progress implies change, but not vice versa; not all changes are progressive. The molecules of oxygen and nitrogen in the air of a room are continuously changing positions, but such changes would not generally be regarded as progressive. The mutation of a gene from a functional allelic state to a nonfunctional one is a change, but definitely not a progressive one.

"Evolution" and "progress" can also be distinguished, although both imply that sustained change has occurred. Evolutionary change is not necessarily progressive. The evolution of a species may lead to its own extinction, a change which is not progressive, at least not for that species. Progress can also occur without evolutionary change. Assume that in a given region of the world the seeds of a certain species are dormant because of a prolonged drought; after a burst of rain the seeds germinate and give origin to a population of plants. This change might be labelled progressive for the species, even though no evolutionary change need to have taken place.

The concept of "direction" implies that a series of changes have occurred which can be arranged in a linear sequence so that elements in the later part of the sequence are further from early elements of the sequence than intermediate elements are. Directional change may be uniform or not, depending whether every later member of the sequence is further displaced than every earlier member, or whether directional change occurs only on the average. Nonuniform or "net" (see below) directional change occurs when the direction of change is not constant; some elements in the sequence may represent a change of direction with respect to the immediately previous elements, but later elements in the sequence are displaced further than earlier ones on the average.

"Directionality" is sometimes equated with "irreversibility" in discussions of evolution: the process of evolution is said to have a direction because it is irreversible. Biological evolution is irreversible (except perhaps in some trivial sense, as when a previously mutated gene mutates back to its former allelic state). Direction, however, implies more than irreversibility. Consider a new pack of cards with each suit arranged from ace to ten, then knave, queen, king, and with the suits arranged in the sequence of spades, clubs, hearts and diamonds. If we shuffle the cards thoroughly, the order of the cards will change, and the changes will be irreversible by shuffling. We may shuffle again and again until the cards are totally worn out, without ever restoring the original sequence. The change of order in the pack of cards is irreversible but not directional.

Directional changes occur in the inorganic as well as the organic world. The second law of thermodynamics, which applies to all processes in nature, describes sequential changes that are directional, and indeed, uniformly directional. Within a closed system, entropy always increases; that is, a closed system passes continuously from less probable to more probable states. The concept of direction applies to what in paleontology are called "evolutionary trends." A trend occurs in a phylogenetic sequence when a feature persistently changes through time in the members of a sequence. Trends are common occurrences in all fossil sequences which are sufficiently long to be called "sustained" (Simpson 1953).

The concept of direction and the concept of progress are distinguishable. Consider the trend in the evolutionary sequence from fish to man towards a gradual reduction in paleontological time of the number of dermal bones in the skull roof; or the trend towards increased molarization in the last premolar which occurred in the phylogeny of the *Equidae* from the early Eocene

(*Hyracotherium*) to the early Oligocene (*Haplohippus*). These trends indeed represent directional change, but it is not obvious that they should be labeled progressive; to do this we would have to agree that the directional change had been for the better in some sense. That is, in order to consider a directional sequence progressive we need to add an evaluation, namely that the condition of the latter members of the sequence represents, according to some standard, a melioration or improvement. The directionality of the sequence can be recognized and accepted without any such evaluation being added. Progress implies directional change, but not vice versa.

Evolution, direction, and progress all imply a historical sequence of events that exhibit a systematic alteration of a property or state of the elements in the sequence. Progress occurs when there is directional change towards a *better* state or condition. The concept of progress, then, contains two elements: one *descriptive*, that directional change has occurred; the other *axiological* (= evaluative), that the change represents betterment or improvement.

The notion of progress requires that a value judgment be made about what is better and what is worse, or what is higher and what is lower, according to some axiological standard. But contrary to the belief of some authors (Ginsberg 1944, Lewontin 1968), the axiological standard of reference need not be a moral one. Moral progress is possible, but not all forms of progress are moral. The evaluation required for progress may be one of better versus worse, or higher versus lower, but not necessarily one of right versus wrong. "Better" may simply mean more efficient, more abundant, or more complex, without connoting any reference to moral values or standards.

One may, then, define progress as "systematic change in a feature belonging to all the members of a sequence in such a way that posterior members of the sequence exhibit an improvement of that feature." More simply it can be defined as "directional change towards the better." Similarly, regress or retrogression is directional change for the worse. The two elements of the definition, namely directional change and improvement according to some standard, are jointly necessary and sufficient for the occurrence of progress.

Directional change, as well as progress, may be observed in sequences that are spatially rather than temporally ordered. Clines are examples of directional change recognized along a spatial dimension. In evolutionary discourse, however, historical sequences are of greatest interest.

Varieties of Progress

In order to seek further clarification of the concept of progress and its application in evolutionary biology, it is necessary to distinguish among various kinds of progress. This can be accomplished according to either one of the two essential elements of the definition. I shall later refer to types of progress differentiated on the basis of axiological standards of reference. Now, I shall make two distinctions that relate to the descriptive element of the definition, i.e., the requirement of directional change. The first distinction takes into account the *continuity* of the direction by distinguishing between "uniform" and "net" progress. The second distinction refers to the *scope* of the sequence considered and differentiates between "general" and "particular" progress.

Uniform progress takes place whenever every later member of the sequence is better than every earlier member of the sequence according to a certain feature. This may be formally stated as follows. Let m_i be the members of the sequence, temporally ordered from 1 to n, and let p_i measure the state of the feature under evaluation. There is uniform progress if it is the case for every m_i and m_j that $p_j > p_i$ for every $j > i$.

Net progress does not require that every member of the sequence be better than all previous members of the sequence and worse than all its successors; it requires only that later members of the sequence be better, *on the average*, than earlier members. Net progress allows for temporary fluctuations of value. Formally, if the members of the sequence, m_i, are linearly arranged over time, net progress occurs whenever the regression (in the sense used in mathematical sequence) of p on time is significantly positive. Some authors have argued that progress has not occurred in evolution because no matter what standard is chosen, fluctuations can always be found in every evolutionary lineage. This argument is valid against the occurrence of uniform progress, but not against the existence of net evolutionary progress.

Notice also that neither uniform nor net progress requires that progress

be unlimited, or that any specified goal will be surpassed if the sequence
continues for a sufficiently long period of time. Progress requires a gradual
improvement in the members of the sequence, but the rate of improvement may
decrease with time. According to the definition given here, it is possible that
the sequence tends asymptotically towards a definite goal, which is
continuously approached but never reached.

The distinction between uniform and net progress is similar but not
identical to the distinction between uniform and perpetual progress proposed by
Broad (1925) and Goudge (1961). Perpetual progress, as defined by Broad,
requires that the maximum of values increase, and the minimum does not
decrease with time. In the formulation given above, Broad's perpetual progress
requires that for every m_i there is at least one m_j ($j > i$) such that $p_j > p_i$.
This definition has the undesirable feature of requiring that the first element
of the sequence be the worst one and the last element the best one. Neither
of these two requirements is made in my definition of net progress. Also the
term "perpetual" has connotations that are undesirable in the discussion of
progress. The distinction between uniform and net progress is made implicitly,
although never formally stated, by Simpson (1949), who applies terms like
"universal," "invariable," constant," and "continuous" to the kind of progress
that I have called uniform (although he also uses these terms with other
meanings).

With respect to the scope of the sequence considered, progress can be
either general or particular. *General progress* is that which occurs in all
historical sequences of a given domain of reality and from the beginning of
the sequences until their end. *Particular progress* is that which occurs in one
or several but not all historical sequences, or that which takes place during
part but not all of the duration of the sequences.

General progress would have occurred in evolution only if there were
some feature or standard according to which progress can be predicated of the
evolution of all life from its origin to the present. If progress is predicated
of only one or several, but not all, lines of evolutionary descent, it is a
particular kind of progress. Progress that embraces only a limited span of
time from the origin of life to the present is also a particular kind of
progress.

Some writers have denied that evolution is progressive on the grounds that not all evolutionary lineages exhibit advance. Some evolutionary lineages, like those leading to certain parasitic forms, are retrogressive by certain standards; and many have lineages have become extinct without issue. These considerations are valid against a claim of general progress, but not against claims of particular forms of progress.

Attempts to Define Progress as a Biological Concept

I have argued that the notion of progress is axiological and, therefore, it cannot be a strictly scientific term: value judgments are not part and parcel of scientific discourse, which is characterized by empirically testable hypotheses and objective descriptions. Some authors have claimed, however, that there are biological criteria of progress that are "objective" and do not involve value judgments. I will briefly review the efforts in this regard of three distinguished biologists: J. M. Thoday, M. Kimura, and J. S. Huxley.

Thoday (1953, 1958) has pointed out the obvious fact that survival is essential to life. Therefore, he argues, progress is the increase in fitness for survival, "provided only that fitness and survival be defined as generally as possible." According to Thoday, fitness must be defined in reference to groups of organisms that can have common descendants; these groups he calls "units of evolution." A unit of evolution is what population geneticists call a Mendelian population; the most inclusive Mendelian population is the species. The fitness of a unit of evolution is defined by Thoday as "the probability that such a unit of evolution will survive for a long period of time, such as 10^8 years, that is to say will have descendants after the lapse of that time." According to Thoday, evolutionary changes, no matter what other results may have been produced, are progressive only if they increase the probability of leaving descendants after long periods of time. He correctly points out that this definition has the advantage of not assuming that progress has in fact occurred, an assumption which vitiates other attempts to define progress as a purely biological concept.

The definition of progress given by Thoday has been criticized because it apparently leads to the paradox that progress is impossible, in fact, that regress is necessary since any group of organisms will be more progressive than any of their descendants. Assume that we are concerned with ascertaining whether progress has occurred in the evolutionary transition from a Cretaceous mammal to its descendants of 100 million (10^8) years later. It is clear that if the present-day mammal species has a probability, P, of having descendants 10^8 years from now, the ancestral mammal species will have a probability no smaller than P of leaving descendants after 2×10^8 years from the time of their existence (Ayala 1969). The probability that the ancestral species will leave descendants 2×10^8 years after their existence will be greater than P if it has any other living descendants besides the present day mammal with which we are comparing it. As Thoday (1970) himself has pointed out, such criticism is mistaken, since it confuses the probability of survival with the fact of survival. The *a priori* probability that a given species will have descendants after a given lapse of time may be smaller then the *a priori* probability that any of its descendants will leave progeny after the same length of time.

There is, however, a legitimate criticism of Thoday's definition of progress, namely, that it is not operationally valid. Suppose that we want to find out whether today's mammal species is more progressive than the Cretaceous ancestor. We should have to estimate, first, the probability that today's mammal will leave descendants after a given long period of time; then, we should have to estimate the same probability for the Cretaceous species. Thoday has enumerated a variety of components which contribute to the fitness of a population as defined by him. These components are adaptation, genetic stability, genetic flexibility, phenotypic flexibility, and the stability of the environment. But it is by no means clear how these components could be quantified nor by what sort of function they could be integrated into a single parameter. In any case, there seems to be no conceivable way in which the appropriate observations and measurements could be made for the ancestral species. Thoday's definition of progress is extremely ingenious, but lacks operational validity. If we accept his definition there seems to be no way in which we could ascertain whether progress has occurred in any one line of descent or in the evolution of life as a whole.

Another attempt to consider evolutionary progress as a purely biological notion has been made by Motoo Kimura (1961), by defining biological progress as an increase in the amount of genetic information stored in the organism.

This information is encoded, at least for the most part, in the DNA of the nucleus. The DNA contains the information which in interaction with the environment directs the development and behavior of the individual. By making certain assumptions, Kimura has estimated the rate at which genetic information accumulates in evolution. He calculates that in the evolution of "higher" organisms genetic information has accumulated from the Cambrian to the present at an average of 0.29 *bits* per generation.

This method of measuring progressive evolution by the accumulation of genetic information is vitiated by several flaws. First, since the average rate of accumulation of information is allegedly constant *per generation*, it follows that organisms with a shorter generation time will have accumulated more information, and therefore are more progressive than organisms with a longer generation time. In the evolution of mammals, moles and bats would necessarily be more progressive than horses, whales and men. A second flaw is that Kimura is not measuring how much genetic information has been accumulated in any given organism. Rather he assumes that genetic information gradually accumulates with time and he then proceeds to estimate the rate at which genetic information could have accumulated. The assumption that more recent organisms have more genetic information and that, therefore, they are more progressive than their ancestors is unwarranted, and completely invalidates Kimura's attempt to measure evolutionary progress. There is, at least at present, no way of measuring the amount of genetic information present in any one organism. Finally, the decision to consider the accumulation of genetic information as progressive requires a value judgment; it is not a biologically compelling notion.

Julian Huxley (1942, 1953) has argued that the biologist should not attempt to define progress *a priori*, but rather he should "proceed inductively to see whether he can or cannot find evidence of a process which can be legitimately called progressive." He believes that evolutionary progress can be defined without any reference to values. Huxley proposes first to investigate the features which mark off the "higher" from the "lower" organisms. Any evolutionary process is considered progressive in which the features that characterize higher organisms are achieved. But Huxley, like Kimura, assumes that progress has in fact occurred, and that certain living organisms, especially man, are more progressive than others. Classifying organisms as "higher" or "lower" requires an evaluation. Huxley has not succeeded in avoiding reference to an axiological standard. The terms that he uses in his various definitions of progress, like "improvement," "general advance," "level of efficiency," etc., are all, in fact, evaluative.

Is Progress a General Property of Evolution?

No attempt to define progress as a purely biological concept has succeeded. This is understandable in view of the analysis of the concept that I developed above. The concept of progress is axiological and, hence, one cannot ascertain whether progress has occurred without first choosing a standard against which progress or improvement will be assessed. Two decisions are required. First, we must choose the objective feature according to which the events of objects are to be ordered. Second, a decision must be made as to what pole of the ordered elements represents improvement. These decisions involve a subjective element, but they should not be arbitrary. Biological knowledge should guide them. There is a criterion by which the validity of a standard of reference can be judged. A standard is valid if it enables us to say illuminating things about the evolution of life. How much of the relevant information is available, and whether the evaluation can be made more or less exactly, should also influence the choice of values.

It is fairly apparent that there is no standard by which *uniform* progress can be said to have taken place in the evolution of life. Changes in direction, slackening, and reversals have occurred in all evolutionary lineages, no matter what features are considered (Simpson 1949, 1953). The question, then, is whether *net* progress has occurred in the evolution of life and in which sense.

The next question is whether there is any criterion of progress by which net progress can be said to be a general feature of evolution; or whether identifiable progress applies only to particular lineages or in particular periods. One conceivable standard of progress is the increase in the amount of genetic information stored in the organisms. Net progress would have occurred if organisms living at a later time would have, on the average, a greater content of genetic information than their ancestors. One difficulty, insuperable at least for the present, is that is no way in which the genetic *information* present in an organism can be measured. We could choose the Shannon-Weaver solution, as Kimura has done, by regarding all the DNA of an organism as a linear sequence of messages made up of groups of three-letter words (the codons) with a four-letter alphabet (the four DNA nucleotides). But the amount of information is not simply related to the amount of DNA, since we know that many DNA sequences are repetitive and even much of the nonrepetitive DNA may not store information in the nucleotide sequence. In any case, what we know about the size of the genome in organisms makes it unlikely that increase of genetic information could be a general feature of the evolution of life.

But the accumulation of the genetic information as a standard of progress can be understood in a different way. Progress can be measured by an increase in the *kinds* of ways in which the information is stored and as an increase in the *number* of different messages encoded. Different species represent different kinds of messages; individuals are messages or units of information. Thus understood, whether an increase in the amount of information has occurred reduces to the question of whether life has diversified and expanded. This has been recognized by Simpson (1949) as the standard by which, what I call, general progress has in fact occurred in the evolution of life. According to Simpson, we can find out about evolution as a whole "tendency for life to expand, to fill in all the available spaces in the livable environments, including those created by the process of that expansion itself."

There are at least four different though related criteria by which the expansion of life can be measured: (1) expansion in the number of *kinds* of organisms, that is, the number of species, (2) expansion in the number of individuals, (3) expansion in the total bulk of living matter, and (4) expansion in the total rate of energy flow. Increases in the number of individuals or of their bulk may be a mixed blessing, as it is the case now for the human species, but they can be a measure of biological success. By any one of the four standards of progress enumerated, it appears that the net progress has been a general feature of the evolution of life.

Reproduction provides organisms with the potentiality to multiply exponentially: each organism is capable of producing, on the average, more than one progeny throughout its lifetime. The actual rate of increase in numbers is a net result of the balance between the rate of births and the rate of deaths of the population. In the absence of environmentally imposed restrictions, that balance is positive; populations have an intrinsic capacity to grow *ad infinitum*. Since the ambit in which life can exist is limited, and since the resources to which a population has access are even more limited, the rate of expansion rapidly decreases to zero, or becomes negative.

The tendency of life to expand encounters constraints of various sorts. The expansion is limited by the environment in at least two ways. First, the supply of resources accessible to the organisms is limited. Second, favorable conditions necessary for multiplication do not always occur. Predators, parasites, and competitors, together with the various parameters of the environment embodied by the term "weather," are the main factors interfering with the multiplication of organisms even when the resources are available. Drastic and secular changes in the weather, as well as geological events, lead at times to vast decreases in the size of some populations and even the whole of life. Because of these constraints the tendency of life to expand has not always succeeded. Nevertheless, it appears certain that life has on the average, expanded throughout most of its history.

About one-and-a-half million species now living have been described and named. Current estimates place the number of living species between five and thirty million, with most of the unidentified species being beetles and other insects. Although it is difficult to estimate the number of plant species which existed in the past (since well-preserved plant fossils are rare), the number of animal species can be roughly estimated. Approximately 150,000 animal species live in the seas today, probably a larger number than the total number that existed in the Cambrian (600 million years ago) when no animal or plant species lived on the land. Life on land began in the Devonian (400 million years ago). The number of animal land species is probably at a maximum now, even if we exclude insects. Insects make up about three-quarters of all known animal species, and about half of all species if plants are included. Insects did not appear until the early Carboniferous, some 350 million years ago. The number of living insect species became larger within the current geological than it was at most, probably all, times in the past. On the whole, it appears that the number of living species is probably greater in recent times (before twentieth century extinctions) than it ever was before, and that, at least on the average, a gradual increase in the number of species has characterized the evolution of life (Simpson 1949, 1953).

The number of species expands by a positive feedback process. The greater the number of species, the greater the number of environments that are created for the new species to exploit. Once there were plants, animals could come into existence, and the animals themselves sustain large numbers of species of other animals that prey on them, as well as of parasites and symbionts. Thomas Huxley likened the expansion of life to the filling of a barrel. First, the barrel is filled with apples until it overflows; then pebbles are added up to the brim; the space between the apples and the pebbles can be packed with sand; water is finally poured until it overflows (see Huxley and Huxley 1947). His point is that with diverse kinds of organisms the environment can be filled in more effectively than with only one kind. But Huxley's analogy neglects one important aspect of life, namely that the space available for occupancy by other species is increased rather than decreased by some additions. A more appropriate analogy would have been that of a balloon or an expanding barrel.

It is difficult to estimate the number of individuals living on the earth today with any reasonable approximation, even if we exclude microorganisms. The mean number of individuals per species has been estimated to be around 2×10^8, but some species like *Drosophila willistoni*, the tropical fruit fly with which I have worked for twenty years, may consist of more than 10^{16} individuals -- and there are more than one million insect species! (Ayala et al. 1972). The number of individuals of *Euphausia superba*, the small krill eaten by whales, may be greater than 10^{20}. It seems certain that there are more individual animals and plants living today and their bulk is greater than it was in the Cambrian. Very likely, they have become greater in recent geological times than they were at most times since the beginning of life. This is more so if we include the large number and enormous bulk of the human population and of all the plants and animals cultivated by man for his own use. Even if we include microorganisms, it is probable that the number of living individuals has increased, on the average, through the evolution of life. An increase in the total bulk of living matter is even more likely than an increase in numbers because larger organisms have generally appeared later in time.

It seems likely that the rate of energy flow has increased in the living world faster than the total bulk of matter. One effect of organisms on the world is to retard the dissipation of energy. Green plants do, indeed, store radiant energy from the sun that would otherwise be converted into heat. The influence of animals goes partially in the opposite direction: the living activities of animals dissipate energy, since their catabolism exceeds their anabolism (Lotka 1945), but they store energy derived from plants that might have otherwise dissipated into heat. Animals provide a new path through which energy can flow and, moreover, their interactions with plants increase the total rate of flow through the system. An analogy can be used to illustrate this outcome. Suppose that a modern highway with three lanes in each direction connects two large cities. The need to accommodate an increase in the rate of travel flow can be accomplished either by adding more lanes to the highway or by increasing the speed at which the traffic moves in the highway. In terms of the "carrying capacity" of the highway, these two approaches appear, at first sight, to work in opposite directions, but together they increase the total flow of traffic on the highway.

Bounded Evolutionary Progress

I have argued above that the concept of progress involves an evaluation of better versus worse relative to some standard of reference. Many standards of reference can be chosen according to which it is possible to measure the evolutionary process of the kind I have called "particular" -- that is, progress that obtains only in certain evolutionary lineages and usually only for a limited span of time. The numerous writers on evolutionary progress have usually proceeded by identifying one or another attribute as the criterion of progress and have then expanded on how progress has occurred in evolution according to the particular standard chosen. These discussions are often enlightening in that they bring about aspects of the evolutionary progress that are particularly meaningful from a certain perspective and enhance our understanding of the process. A common deficiency in some of these discussions is the stated or implicit conviction that *the* criterion of progress has been discovered, often accompanied by a lack of awareness that progress is a value-laden concept rather than a strictly scientific one.

I shall now mention some of the criteria that have been the subjects of thoughtful discussion on evolutionary progress. I will then, by way of illustration, deal in somewhat greater detail with one specific criterion of evolutionary progress: advances in the ability of organisms to obtain and process information about the state of the environment.

Simpson (1949) has examined several criteria according to which evolutionary progress can be recognized in particular sequences. These criteria include dominance, invasion of new environments, replacement, improvement in adaptation, adaptability and possibility of further progress, increased specialization, control over the environment, increased structural complexity, increase in general energy or level of vital processes, and increase in the range and variety of adjustments to the environment. For each of these criteria Simpson has shown in which evolutionary sequences, and for how long, progress has taken place.

Bernhard Rensch (1947) and Julian Huxley (1942, 1953) have examined other lists of characteristics which can be used as standards of particular forms of progress. Ledyard Stebbins (1969) has written a provocative essay proposing a law of "conservation of organization" that accounts for evolutionary progress as a small bias toward increased complexity of organization. I (Ayala 1974, 1982) have examined elsewhere in some detail the increase in the ability of organisms to obtain and process information about the environment, as a criterion of progress that is particularly relevant to the evolution of man; among the differences that mark off humans from all other animals, perhaps the most fundamental is the human's greatly developed ability to perceive the environment and to react flexibly to it. George Williams (1966) has examined, mostly critically, several criteria of progress. Two brief but incisive discussions of the concept of progress can be found in G. J. Herrick (1956) and Theodosius Dobzhansky (1970). A philosophical study of the concept of progress has been made by T. A. Goudge (1961).

There is no need to examine here all the standards of progress which have been formulated by the authors just mentioned, nor to explore additional criteria. Writings about biological progress have involved much disputation concerning (1) whether the notion of progress belongs in the realm of scientific discourse, (2) what criterion of progress is "best," and (3) whether progress has indeed taken place in the evolution of life.

These controversies can be solved once the notion of progress is clearly established. First, the concept of progress involves an evaluation of good versus bad, or of better versus worse. The choice of a standard by which to evaluate organisms or their features is to a certain extent subjective. However, once a standard of progress has been chosen, decisions concerning whether progress has occurred in the living world, and what organisms are more or less progressive, can be made following the usual standards and methods of scientific discourse. Second, there is no standard of progress that is "best" in the abstract or for all purposes. The validity of any one criterion of progress depends on whether the use of that standard leads to meaningful statements concerning the evolution of life. Which standard or standards are preferable depends on the particular context or purpose of the discussion. Third, the distinction between uniform and net progress makes it possible to recognize the occurrence of biological progress even though every member of a sequence or of a group of organisms may not always be more progressive than every previous member of the sequence or than every member of some other group of organisms. Fourth, the distinction between general and particular progress allows one to recognize progress that may have occurred in particular groups of organisms, or during limited periods in the evolution of life, but not in all of them.

Human Consciousness: Climax of One Kind of Progress

Once one realizes that recognition of progress is only possible after a value judgment has been made as to which will be the standard against which progress is to be measured (and hence, that there is not *a* standard of progress, or one that is best for all purposes), it becomes possible to seek standards of progress that may yield valuable insights into the study of the evolution of life.

I shall now, by way of illustration, discuss progress according to a particular standard of reference: the ability of an organism to obtain and process information about the environment. I can see two reasons that make this criterion of progress especially meaningful (although not, I reiterate, *the*

most meaningful, because no criterion exists that is best for all purposes).
First, the ability to obtain information about the environment and to react
accordingly, is an important adaptation, because it allows the organism to seek
out suitable environments and resources and to avoid unsuitable ones. Second,
because the ability to perceive the environment, and to integrate, coordinate,
and react flexibly to what is perceived, has attained its highest development in
mankind. This incomparable advancement is perhaps the most fundamental
characteristic that sets apart *Homo sapiens* from all other animals. Symbolic
language, complex social organization, control over the environment, the ability
to envisage future states and to work towards them, values and ethics are
developments made possible by man's greatly developed capacity to obtain and
organize information about the state of the environment. This capacity has
ushered in mankind's new mode of adaptation. Whereas other organisms
become genetically adapted to their environments, humans create environments
to fit their genes. It is thus that mankind has spread over the whole planet
in spite of its physiological dependence on a tropical or subtropical climate.

Increased ability to gather and process information about the environment
is sometimes expressed as evolution towards "independence from the
environment." This latter expression is misleading. No organism can be truly
independent of the environment. The evolutionary sequence fish to amphibian
to reptile allegedly provides an example of evolution toward independence from
an aqueous environment. Reptiles, birds, and mammals are indeed free of the
need for water as an external living medium, but their lives depend on the
conditions of the land. They have not become independent of the
environment, but have rather exchanged dependence of one environment for
dependence on another.

The notion of "control over the environment" also has been associated
with the ability to gather and use information about the state of the
environment. However, true control over the environment occurs to any
substantial extent only in the human species. All organisms interact with the
environment, but they do not control it. Burrowing a hole in the ground or
building a nest in a tree, like the construction of a beehive or a beaver dam,
does not represent control over the environment except in a trivial sense.
The ability to control the environment started with the australopithecines, the
first group of organisms which may be called human: some were able to

produce devices to manipulate the environment in the form of rudimentary pebble and bone tools. The ability to obtain and process information about the conditions of the environment does not provide control over the environment but rather it enables the organisms to avoid unsuitable environments and to seek suitable ones. It has developed in many organisms because it is a useful adaptation.

Some selective interaction with the environment occurs in all organisms. The cell membrane of a bacterium permits certain molecules but not others to enter the cell. Selective molecular exchange occurs also in the inorganic world; but this can hardly be called a form of information processing. Certain bacteria when placed on a agar plate move about in a zig-zag pattern, which is almost certainly random. The most rudimentary ability to gather and process information about the environment may be found in certain single-celled eukaryotes (= organisms with a true nucleus). A *Paramecium* follows a sinuous path as it swims, ingesting the bacteria that it encounters. Whenever it meets unfavorable conditions, like unsuitable acidity or salinity in the water, the *Paramecium* checks its advance, turns and starts in a new direction. Its reaction is purely negative. The *Paramecium* apparently does not seek its food or a favorable environment, but simply avoids unsuitable conditions.

Euglena, also a single-cell organism, exhibits a somewhat greater ability to process information about the environment. *Euglena* has a light-sensitive spot by means of which it can orient itself towards the light. *Euglena*'s motions are directional; it not only avoids unsuitable environments but it actively seeks suitable ones. An amoeba represents further progress in the same direction; it reacts to light by moving away from it, and also actively pursues food particles.

An increase in the ability to gather and process information about the environment is not a general characteristic of the evolution of life. Progress has occurred in certain evolutionary lines but not in others. Today's bacteria are not more progressive by this criterion than their ancestors of three billion years ago. In many evolutionary sequences some very limited progress took place in the very early stages, without further progress through the rest of their history. In general, animals are more advanced than plants; vertebrates are more advanced than invertebrates; mammals are more advanced than reptiles, which are more advanced than fish. The most advanced organism by this criterion is doubtless the human species.

The ability to obtain and to process information about the environment has progressed little in the plant kingdom. Plants generally react to light and to gravity. The geotropism is positive in the root, but negative in the stem. Plants also grow towards the light; some plants like the sunflower have parts which follow the course of the sun through its daily cycle. Another tropism in plants is the tendency of roots to grow towards water. The response to gravity, to water, and to light is basically due to differential growth rates; a greater elongation of cells takes place on one side of the root or stem than on the other side. Gradients of light, gravity or moisture are the clues which guide these tropism. Some plants react also to tactile stimuli. Tendrils twine around what they touch; *Mimosa* and carnivorous plants like the Venus flytrap (*Dionaea*) have leaves which close upon being touched.

The ability to obtain and process information about the environment is mediated in multicellular animals by the nervous system. All major groups of animals, except the sponges, have nervous systems. The simplest nervous system occurs in coelenterate hydras, corals and jellyfishes. Each tentacle of a jellyfish reacts only if it is individually and directly stimulated. There is no coordination of the information gathered by different parts of the animal. Moreover, jellyfishes are unable to learn from experience.

A limited form of coordinated behavior occurs in the echinoderms which comprise the starfishes and sea urchins. Whereas coelenterates possess only an undifferentiated nerve net, echinoderms possess a nerve net, a nerve ring, and radial nerve cords. When the appropriate stimulus is encountered, a starfish reacts with direct and unified actions of the whole body.

The most primitive form of a brain occurs in certain organisms like planarian flatworms, which also have numerous sensory cells and eyes without lenses. The information gathered in these sensory cells and organs is processed and coordinated by the central nervous system and the rudimentary brain; a planarian worm is capable of some variability of responses and of some simple learning. That is, the same stimuli will not necessarily always produce the same response.

Planarian flatworms have progressed farther than starfishes in the ability to gather and process information about the environment, and the starfishes have progressed farther than sea anenomes and other coelenterates. But none of these organisms has gone very far by this criterion of progress. The most progressive groups of organisms among the invertebrates are the cephalopods and arthropods, but the vertebrates have progressed much farther than any invertebrates.

Among the ancestors of both the arthropods and the vertebrates, there were organisms that, like the sponges, lacked a nervous system. These ancestors evolved through a stage with only a simple network, whereas later stages developed a central nervous system and eventually a rudimentary brain. With further development of the central nervous system and of the brain, the ability to obtain and process information from the outside progressed much farther. The arthropods, which include the insects, have complex forms of behavior. Precise visual, chemical and acoustic signals are obtained and processed by many arthropods, particularly in their search for food and in their selection of mates.

Vertebrates are generally able to obtain and process much more complicated signals and to produce a much greater variety of responses than the arthropods. The vertebrate brain has an enormous number of associative neurons with an extremely complex arrangement. Among the vertebrates, progress in the ability to deal with environmental information is correlated with increase in the size of the cerebral hemispheres and with the appearance and development of the "neopallium." The neopallium is involved in association and coordination of all kinds of impulses from all receptors and brain centers. The larger brain of vertebrates, compared to that of invertebrates, permits them also to have a large amount of neurons involved in information storage or memory. The neopallium appeared first in the reptiles. In the mammals it has expanded to become the cerebral cortex, which covers most of the cerebral hemispheres. The cerebral cortex in humans is particularly large, compressed over the hemispheres in a complex pattern of folds and turns. When organisms are measured by their ability to process and obtain information about the environment, mankind is, indeed, the most progressive organism on earth.

I would once more reiterate that there is nothing in the evolutionary process which makes the criterion of progress I have just followed best or more objective than others. It may be useful because it illuminates certain features of the evolution of life. Other criteria may help to discern other features of evolution, and thus be worth examining. Particular organisms will appear more or less progressive depending on the standard that is used to evaluate progress. Mankind is not the most progressive species by many criteria. By some standards, humans are among the bottom rungs of the ladder of life, for example, in the ability to synthesize their own biological materials from inorganic resources.

It may be properly questioned whether anything is gained by speaking of evolutionary *progress* rather than of evolutionary advancement or of directional change. The term "advancement" also involves an evaluation and would therefore be subject to the same pitfalls as "progress" (although it seems to elicit weaker emotional discharges than progress does). "Directional change," however, is not an axiological concept and, thus, it may be treated as other strictly scientific terms.

If the term "progress" were to be completely obliterated from scientific discourse, I would be quite pleased; but it seems to me unlikely to happen. The notion of progress seems to be irrevocably ingrained among the thinking categories of modern man and, hence, likely to continue being used in biology, particularly in reference to the evolutionary progress. I have, therefore, attempted to clarify the concept in order to demythologize it. I have argued that "progressive" is an evaluative term that demands a subjective commitment to a particular standard of value. Awareness of this makes it possible to speak of progress in evolution without implying the conclusion that humans are the most progressive organisms. By some biologically meaningful standards of progress they are not.

References

Ayala, F. J. 1969. An evolutionary dilemma: Fitness of genotypes versus fitness in populations. *Canadian Journal of Genetics and Cytology* 11:439-56.

Ayala, F. J. 1974. The concept of biological progress. In *Studies in the philosophy of biology*, ed. F. J. Ayala and T. Dobzhansky. Berkeley: University of California Press.

Ayala, F. J. 1982. The evolutionary concept of progress. In *Progress and its discontents*, ed. G. A. Almond, M. Chodorow and R. H. Pearce. Berkeley: University of California Press.

Ayala, F. J., J. R. Powell, M. L. Tracey, C. A. Mourao, and S. Perez-Salas. 1972. Enzyme variability in the *Drosophila willistoni* group. IV. Genic variation in natural populations of *Drosophila willistoni*. *Genetics* 70:113-39.

Broad, C. D. 1925. *The mind and its place in nature*. London: Kegan Paul.

Dobzhansky, T. 1970. *Genetics of the evolutionary process*. New York and London: Columbia University Press.

Ginsberg, M. 1944. Moral progress. Frazer Lecture at the University of Glasgow. Glasgow: Glasgow University Press.

Goudge, T. A. 1961. *The ascent of life*. Toronto: University of Toronto Press.

Herrick, G. J. 1956. *The evolution of human nature*. Austin: The University of Texas Press.

Huxley, J. S. 1942. *Evolution, the modern synthesis*. New York: Harper.

Huxley, J. S. 1953. *Evolution in action*. New York: Harper.

Huxley, T. H., and J. S. Huxley. 1947. *Touchstone for ethics*. New York: Harper.

Kimura, M. 1961. Natural selection as the process of accumulating genetic information in adaptive evolution. *Genetical Research* 2:127-40.

Lewontin, R. C. 1968. The concept of evolution. In *International Encyclopedia of the Social Sciences*, vol. 5, ed. D. L. Sills. London and New York: Macmillan Co. and Free Press.

Lotka, A. J. 1945. The law of evolution as a maximal principle. *Human Biology* 17:167-94.

Lovejoy, A. O. 1936. *The great chain of being*. Cambridge: Harvard University Press.

Rensch, B. 1947. *Evolution above the species level.* New York: Columbia University Press.

Simpson, G. G. 1949. *The meaning of evolution*. New Haven: Yale University Press.

Simpson, G. G. 1953. *The major features of evolution*. New York: Columbia University Press.

Stebbins, G. L. 1969. *The basis of progressive evolution*. Chapel Hill: University of North Carolina Press.

Thoday, J. M. 1953. Components of fitness. Symposia of the Society for the Study of Experimental Biology, 7 (Evolution):96-113.

Thoday, J. M. 1958. Natural Selection and biological progress. In *A century of Darwin*, ed. S. A. Barnet. London: Allen and Unwin.

Thoday, J. M. 1970. Genotype versus population fitness. *Canadian Journal of Genetics and Cytology* 12.

Williams, G. C. 1966. *Adaptation and natural selection*. Princeton: Princeton University Press.

Molecules to Men: Evolutionary Biology and Thoughts of Progress

Michael Ruse

"To deride the hope of progress is the ultimate fatuity, the last word in poverty of spirit and meanness of mind. There is no need to be dismayed by the fact that we cannot yet envisage a definitive solution of our problems, a resting-place beyond which we need not try to go." ---Medawar 1972, p. 127

The hundredth anniversary of Charles Darwin's death fell in 1982. As is their wont, scholars from the world over gathered in meetings and conferences and workshops to mark the occasion, looking both at the historical Darwin and at his legacy for us today (Wassersug and Rose 1984). Without doubt, the most prestigious and comprehensive meeting was in Darwin's own university town of Cambridge, where historians, philosophers and biologists of all kinds came together to pay their respects and to carry on the work of the *Origin of Species.*

In due time, the papers of the Cambridge conference were brought together and published (Bendall 1983). And, there's the rub. Or rather, in the title of the collection lies the spur for this discussion now. That which was bestowed -- *Evolution from Molecules to Men* -- strikes any good Darwinian as anomalous, to put it mildly. It is well known that there is, in evolutionary biology, a bad trend and a good trend. (See Ruse 1986 for all the details.) The bad trend, which goes back at least to Darwin's contemporary, Herbert Spencer, tries to turn evolution into a progressive doctrine -- from least to most, from worst to best. The good trend, whose most notable exponent was Darwin himself, sees evolution as a directionless process, going nowhere rather slowly. Hence, that the closest thing to an official commemoration of Darwin should put him in the wrong camp is insensitive, to say the least.

97

And yet, although this is what I thought when the volume appeared, and although several of the organizers and contributors have told me of their own irritation and/or embarrassment on this score, I am no longer quite so sure that the collection was radically mistitled. At least, let me put things this way. Having now spent considerable time with some of the biological contributors and having pored over their works, and having gone back to evolution's history -- especially to the writings of Darwin -- I am no longer so confident about the old dichotomy between good and bad biology, where the forces of light are against progression and the forces of darkness are for it.

I am not now about to argue for progress in biology. Apart from anything else, I am an historian and philosopher and not a biologist. But I still believe that at a reasonably empirical level, any notion of progress is hard to objectify and instantiate (Hull this volume; Maynard Smith this volume). This is not to say that, in principle, it cannot be done. Indeed, if life started with individual molecules, there really had to be some development of some kind. But, once life gets going, to capture an objective sense of progress, actual attempts generally prove less than compelling. For instance, to take but one possible line of inquiry, if one equates progress with cellular DNA content, then it is indeed true that, confirming our intuitions, humans come out well ahead of bacteria like *E. coli*. On the other hand, we lag pitifully behind certain reptiles and plants. You can, of course, respond by adding qualifications -- epicycles on deferents -- but before long your position looks suspiciously Ptolemaic. You may be right, but you are hardly "reasonably empirical."

What this all leads one to suspect is that thoughts of progress are more put upon science by its practitioners than thrust upon the practitioners by the world "out there." At least, this is the hypothesis I want to explore in this discussion. But as I begin, do let me make one thing clear. Although I have some strong views on progress in general and biological progress in particular, my aim here is not critical. This is not an *exposé* of the mystical, sociopolitical yearnings of the average evolutionist, past or present. I have long argued that evolutionary theory is a good tough scientific theory and that its practitioners command respect. Rather, what I want to see is precisely the true nature of evolutionary thought, and how it is constructed. I hope that such an inquiry as this will throw light on both biology and philosophy.

I begin by looking briefly at the history of evolutionary biology. Then, I turn to contemporary thought. Finally, I draw out some implications and conclusions.

Spencer and Darwin on Progress

For reasons which I shall make clear, my historical cutoff date is 1966. Taken as a whole, this gives us an historical span of about 200 years, for it was in the middle of the eighteenth century that people like Buffon began to doubt the fixity of species in a serious and systematic manner (Bowler 1984; Ruse 1979). However, to get on with my story, I shall collapse the first hundred years into a very few words.

What one can say with a fair amount of authority is this. First, if one uses the definition of the great historian of ideas, Arthur Lovejoy, that progress is a "tendency inherent in nature or in man to pass through a regular sequence of stages of development in past, present and future, the later stages being -- with perhaps occasional retardations or minor retrogressions -- superior to the earlier" (Lovejoy and Boas 1935), then (like evolution) the doctrine of progress first really caught fire in the eighteenth century. There are certainly intimations of progress-like doctrines in earlier writers, but it was such enthusiasts as Turgot, Adam Smith, and above all Condorcet that made the idea an ideology (Almond et al. 1982).

Second, the earliest evolutionists were unambiguously progressionist in their thinking about the organic world. Jean Baptiste de Lamarck is a paradigm. He saw all organisms as being on a ladder, or rather an escalator, with the humblest at the bottom and the proudest, namely humans, at the top. (Actually, he believed in two ladders, one for animals and one for plants.) For Lamarck, evolution is all a matter of progress up the ladder, from the moment of spontaneous generation, when worm-like "monads" are formed, until finally one's ancestors turn from orangutans into humans (Lamarck 1809).

Third, there are strong connections between the first point and the second. It was not chance that started evolutionism in the eighteenth century. Early evolutionists unambiguously saw their beliefs as the organic manifestation of what is elsewhere a socio-politico-economic process. Lamarck shows this, both from his writings and from what we know of his life (Jordanova 1984).

Coming down into the nineteenth century, there are the same close connections between progress and evolutionism. In Britain, for instance, the first fully articulated evolutionary system was that of the Scottish businessman Robert Chambers. In his *Vestiges of the Natural History of Creation*, published anonymously in 1844, he argues for a total monad-to-man picture of the organic world. According to Chambers, the key to evolutionary change lies in embryology. He had been reading his Germans, like Karl Ernst von Baer. Apparently, the longer an organism is in development, in the egg or womb particularly, the more advanced it will be. As it happens, every now and then an organism, for instance a fox, will be prevented from giving birth at its usual time; later, it will give birth to another, more progressed offspring, such as a dog. Ultimately, and taking an overall perspective, one goes from primitive forms (which were spontaneously generated by the action of heat and electricity) to humans.

Because he kept his identity a secret, it is not easy to link Chambers's science with his general views. However, both from internal evidence and from what we know of his life -- he and his brother William were highly successful publishers of a weekly popular newspaper -- his science was at one with his social beliefs (Chambers 1872). And this, I suspect, was true also of the two central evolutionists of the mid-Victorian period, Herbert Spencer and Charles Darwin. In both cases, what we get is not so much science on one side and social beliefs on the other, but rather package deals of both science and social beliefs (Greene 1981).

But what of Spencer and Darwin? Do we find progress on the one side and nonprogress on the other, and is the key distinguishing factor (as is usually claimed) that Darwin alone took seriously the totally nonprogressive mechanism of natural selection? As with so many "well-known" facts, this is only partially true. With Spencer, the traditional account is fairly accurate. With Darwin, things are not quite what they are usually taken to be.

Spencer unambiguously endorses progress, in the biological world as elsewhere. Indeed, for him it is all part and parcel of one general law, of the simple going to the complex, of the homogeneous turning into the heterogeneous (Spencer 1857). In the biological realm, Spencer has (to our eyes, at least) a rather odd mixture of the old and the new. He starts (as does Darwin) with the ever-present Malthusian pressures brought on by potential population increase. This he takes to be the spur for a kind of Lamarckian process of change through need and through the inheritance of acquired characters -- always for Spencer the prime cause of evolutionary change. The overall process makes sense, it evinces an upward climb, because the more you change the more complex you get, and the less prone you are to reproduction. Ultimately with humans you reach the highest point, where already you see the Malthusian pressures easing off, and the struggle no longer so effective (or making necessary further change) (Spencer 1852).

Why should there be this falling away of reproductive power? For Spencer, it was an a priori necessity. As things get more complex, this obviously translates itself into greater brain power, and thus there is clearly less of life's vital nutrients left to contribute to reproduction. However, always ready to back up a nice speculation with a little anecdotal evidence, Spencer pointed out that in real life there seems to be this trade off between sex and the brain. Apparently it all revolves around the fact that what one puts into reproduction takes away from the individual. For example, undue production of sperm cells in man leads first to headaches: "this is followed by stupidity; should the disorder continue, imbecility supervenes, ending occasionally in insanity" (1852, p. 493).

Also it was worth pointing out that whereas herrings have small brains and lots of offspring, humans have large brains and few offspring. In humans taken alone, also, we get life's law at work. The Irish have many children whereas the Scots have but few, and we all know about their relative intellectual abilities. Evolution thus reaches its apotheosis with *Homo britannicus*. Remarkably, even Spencer had the modesty to forbear stressing that he himself was a lifelong bachelor, with no offspring at all.

Turning now to Darwin, we find that, as just mentioned, he too made the Malthusian struggle his starting point. However, for him this pointed primarily in the direction of differential survival and reproduction -- natural selection. Although Darwin was always a Lamarckian -- at times more so than at others -- as a force for change, the inheritance of acquired characteristics was never more than a variation on the main theme (Young 1969). Given that selection was so important for Darwin, since (for all of his ignorance about the cause of heredity) he was always adamant that the variations on which selection works

-- the "raw stuff" of evolution -- are random with respect to the needs of their possessors, it would seem that his views about progression would have to represent a fundamental break with those of Spencer and all who had gone before. (Actually, Darwin had done his major creative evolutionary work before either Chambers or Spencer -- although he delayed publishing until after either of them.) And, there is indeed evidence, internal and external, for such a nonprogressionist reading of Darwin. Most particularly, his root metaphor was of a tree of life (originally a coral of life), not of a ladder or a chain. This means, given the branching, at one level you really cannot get a single progressive upward drive.

Moreover, Darwin himself saw selection and its variants in a nonprogressive way. From the beginning he decried putting one species (us) above all others (Darwin 1960). And, most famously, when he saw the criticisms others made of Chambers he cautioned himself never to use the terms "higher" and "lower." Externally also, apart from the pessimistic influence of Malthus (who wrote explicitly against the optimism of people like Condorcet), there was the fact that Darwin was an intellectual disciple of the geologist Charles Lyell, whose uniformitarian geology was the epitome of nondirectionalism (Rudwick 1969).

Yet, even as one makes the case against progressionism, there are straws being blown the other way. For all of his uniformitarianism, Lyell (1830-33) made humans a special case, calling for direct divine intervention. Darwin's mentor was telling him that some organisms are more equal than others, a point with which other uniformitarians like John Herschel, whose understanding of the organic world was unambiguously progressive, fully agreed (Cannon 1961). Moreover, many of those whom Darwin knew or read -- with sympathy or even enthusiasm -- were prophets of progressionism. Thomas Carlyle, a close intimate of Darwin in the late 1830s, was one (Houghton 1957). (Later, Carlyle became bitterly pessimistic.) Another whose doctrines and works Darwin knew well was the eighteenth century economist Adam Smith (Schweber 1977, 1980). In other words, were Darwin rigidly nonprogressionist, he would have had to have been steering hard against a social current in which he was deeply immersed.

Furthermore, turning back to Darwin's writings, there were always some incredibly progressionist sentiments. Most notorious are the closing lines of

the *Origin*, which go right back to Darwin's earliest writings, namely his
Sketch of 1842 and his *Essay* of 1844 (Darwin and Wallace 1958).

> Thus, from the war of nature, from famine and death, the most exalted
> object which we are capable of conceiving, namely, the production of the
> higher animals, directly follows. There is grandeur in this view of life,
> with its several powers, having been originally breathed into a few forms
> or into one; and that, whilst this planet has gone cycling on according to
> the fixed law of gravity, from so simple a beginning endless forms most
> beautiful and most wonderful have been, and are being, evolved. (Darwin
> 1959, p. 490)

It is hard to imagine that these words are simply a rhetorical flourish.

My sense is that for many years after he became an evolutionist, there
were tensions in Darwin's mind between such progressive sentiments as these
just noted and all the objections (detailed above) of such a theory as Darwin's
to any kind of progressionism. The resolution seems to have come in the
mid-1850s, although the announcement was not really made until 1861, in the
third edition of the *Origin* (1959), by which time Darwin was no doubt
emboldened by the generally progressivist interpretation his favorable readers
gave to his ideas, no matter what he really said. Finally, therefore, Darwin
gave in and went along with the progressionist current (Ospovat 1981; Richards
this volume).

Essentially, his view of progress had two prongs. On the one hand, there
is a peculiarly Darwinian notion of "competitive highness." Natural selection
leads to adaptation, but the latter is a relative notion. You are only as good
as your last victory. If you come into contact with other organisms, then the
struggle is on and the best one wins. However, as one might expect, the more
tested an adaptation, the better it will generally perform.

> Natural selection acts, as we have seen, exclusively by the preservation
> and accumulation of variations, which are beneficial under the organic
> and inorganic conditions of life to which each creature is at each
> successive period exposed. The ultimate result will be that each creature
> will tend to become more and more improved in relation to its conditions
> of life. This improvement will, I think, inevitably lead to the gradual
> advancement of the organization of the greater number of living beings
> throughout the world. (Darwin 1959, p. 221, *382.2:c-4:c*).

On the other hand, Darwin adopted some of the most sophisticated scientific thinking of his time, which argued that a tendency toward divergence and specialization is a mark of progress. Here, Darwin starts to sound remarkably like Spencer on heterogeneity, which is hardly surprising since they drew on the same sources. What Darwin did was to combine these two prongs, arguing that (in general) what competitive highness will lead to is precisely specialization. Hence, progress does indeed come out of a selection-driven evolutionary process.

> If we look at the differentiation and specialization of the several organs of each being when adult (and this will include the advancement of the brain for intellectual purposes) as the best standard of highness of organization, natural selection clearly leads towards highness; for all physiologists admit that the specialization of organs, inasmuch as they perform in this state their functions better, is an advantage to each being; and hence the accumulation of variations tending towards specialization is within the scope of natural selection. (Darwin 1959, p. 222, *382.11:c*)

Once he had got this, then Darwin was off and running. By the time he had come to the *Descent of Man* (1871), he had absorbed all of the Victorian ideals with respect to the virtues of white, Anglo-Saxon, capitalist, Protestant males -- and they come right back to us on virtually every page. Do not worry too much about the Irish: Although they outbreed the Scots, they care less for their children and so will lose in the long run (shades of *r* versus *K* selection!). Do value the moneyed classes: They provide leisure for the talented members of society (like the grandson of Josiah Wedgwood?). Treat women well: Their minds are like children, but they do understand the affairs of the heart (as one learns from Dickens, and other popular novelists). And so on and so forth, right through to various anthropological reports on the sexual behaviors of the blacks.

People who deny that Darwin was a progressionist -- and I was one of them (Ruse 1979) -- are just plain wrong. He believed in molecules to men (and by "men" I mean "men") no less than did his predecessors and contemporaries. The die had been set, within a selectionist view of the world as well as without.

Progress after Darwin

Now, at this point, I shall start to speed up the clock again to bring us down to our own time. As just about every historian of the period has noted, from Darwin's German supporter Ernst Haeckel down to Darwin's fellow Englishman, Julian Huxley, ideas of progress have ridden high (Bowler 1984; Mayr 1982). Usually historians have accompanied this realization with a fair amount of bemoaning the ignorance of the real Darwin; but, as we have just seen, people were in fact treading in the footsteps of the master. Apparently, you can get inertia no less in the world of ideas than you can in the world of organisms.

Of course, I do not want to say that everyone believed in absolutely the same notion of progress as did Darwin or other of his contemporaries -- although you should by now be aware that I do not see the great difference between Darwin and Spencer that has often been assumed (with respect to progress, that is). Some fairly predictable differences reflect national sensibilities. Although Haeckel (1883) trumpeted himself as Darwin's great champion on the Continent, when it came to the apotheosis of progression you will perhaps not be surprised to learn that the top organisms spoke real German rather than some bastard dialect from certain North Atlantic offshore islands.

Also it is true that not every evolutionist in the hundred years after the *Origin* was an out-and-out progressionist. For instance, the American paleontologist Alpheus Hyatt is noted for his theory of racial senescence (Gould 1977b; Richardson and Kane, this volume). Apparently groups of organisms have a youth, a maturity, and then decline into senile old age. Yet, with exceptions proving rules, it is worth noting in this context that inasmuch as people were non- or antiprogressionists, this tended not to stem from any new or revised empirical discoveries or theoretical constructs. Biologists tended rather to keep in tune with general thinking. It is well recorded that by the beginning of this century -- with the expansion of colonialism, the rise of militarism, and the failure to solve social problems -- many intellectuals were starting to lose faith in progress (Almond et al. 1982). Obviously this loss was made more complete in this century by two world wars and the perceived failure of such new beginnings as the Bolshevik Revolution. When biologists pulled back they were doing no more than anyone else.

Yet, even while acknowledging this point towards an understanding of today's thinking, it is worth noting that if you wanted to find the optimists about progress in any group, the evolutionary biologists were often a good place to start. Even if they had doubts about present times, they had pride in the past and suggestions for a brighter future. Instructive in this respect is Sir Ronald Fisher, one whose significance within the history of evolutionary theory is indubitable, for he, with Sewall Wright and J. B. S. Haldane, brought Mendelism to Darwinism. Although Fisher said little about the general course of evolution, it is fairly clear that he saw it as generally progressive, both relatively (as in the Darwinian notion of comparative success) and absolutely (as culminating in humans). Certainly there are strong hints, both from the text of *The Genetical Theory of Natural Selection* (1930) and from outside comments (Bennett 1983; Fisher Box 1978), that the fundamental theorem of natural selection is to be seen as a progression-causing mechanism (or description of such a mechanism) in some sense. It is undeniable that we have the notion of comparative success. That is what the theorem is all about. Additionally, for all the disclaimers, there is an odor of absolute worth. Why else liken the theorem to the second law of thermodynamics?

Interestingly, in this context, the disanalogy that Fisher sees between the two laws makes this point:

> "[E]ntropy changes lead to a progressive disorganization of the physical world, at least from the human standpoint of the utilization of energy, while evolutionary changes are generally recognized as producing progressively higher organization in the organic world." (1930, p. 40)

Moreover, what cannot be doubted is that Fisher thought the fundamental theorem had led to a pretty remarkable animal, namely ourselves. Most of his waking thoughts, not to mention a goodly chunk of the *Genetical Theory*, was given over to the subject. And Fisher was worried. Thanks to a heavy reading (aloud) program which he and his wife had completed during the First World War, Fisher realized that all worthwhile civilizations go into decline -- and he thought he knew the reason why. Advanced though humans may be, as soon as they get themselves into positions of power and affluence, they start to reduce their birth rates. Hence, before long, only the biologically inferior are reproducing at a normal rate, and decline and disaster follow not far behind.

Fortunately, Fisher thought he knew the cure: positive eugenics. What is needed is that the upper classes have more children, and he himself obliged to the extent of having a family of six. Yet, at this point, I am not really interested in the fine points (or even virtues) of such beliefs and/or actions. Here, I want simply to use Fisher to stress two points. On the one hand, despite all the counterevidence around him, be it from his reading or from the War, note how Fisher the biologist remained convinced of the possibility and opportunity of progress. There has been progress in the past, and despite the obstacles, thanks to our selection-produced powers of reason, progress is possible in the future. Such optimism must be explained.

On the other hand, note how moving across the Mendelian divide made little or no difference whatsoever to thoughts of progress. You might think that the nondirectedness of mutations would make thoughts of progress obsolete. This was not so. Again, therefore, we see an obsession which calls for understanding. In this context, incidentally, it is amusing to compare Fisher with his contemporary, the American paleontologist Henry Fairfield Osborn. The latter never accepted the evolutionary implications of Mendelism. But, what really separated them had little to do with biology. Rather, whereas the upper-class Englishman was worried by the threat posed by the fecundity of the working classes, the upper-class American was worried by the threat posed by the fecundity of the immigrant classes, especially the Jews (Osborn 1929; Gould 1982a).

And so we come to the 1930s and 1940s and to the founding of the synthetic theory. One of the chief architects, Julian Huxley (1942) in England, has already been noted as a progressionist, and it is well known that the same holds true of the chief architect in America, Theodosius Dobzhansky (1962, 1967). Huxley particularly brought progressivism right into his theorizing -- but Dobzhansky was never far behind. There is something of a tendency today, especially in evolutionary circles, to regard such open enthusiasm with some embarrassment. It is felt that, talented though the two men undoubtedly were, they had soft intellectual underbellies. Dobzhansky had yearnings for the Russian Orthodox faith of his youth, and Huxley in his "secular humanism" was trying to compensate for the aching void of his family's agnostic tradition.

How else can one explain their shared fuzzy-headed delight in the wild progressive speculations of that heretical Jesuit theorist, Teilhard de Chardin (1959)?

Yet while all of this may be true, it must be noted that in their progressionism Huxley and Dobzhansky were not alone. In America (to take one instance), three other names are always linked with Dobzhansky's in talk of the synthetic theory -- Ernst Mayr (1942, 1963), the ornithologist-systematist; George G. Simpson (1944, 1953), the paleontologist; and G. Ledyard Stebbins (1950), the botanist. All three were or are progressionists, seeing humans as the culmination of the evolutionary process (and with extrascientific enthusiasm for progress generally). It is true that they have tended to be a little less flagrant than Huxley and Dobzhansky. The usual method is to write a "serious" book and then a second "popular" book with all of the progressionism (e.g., Simpson 1963). But the two are rarely that distinct.

Stebbins (1969), for instance, has what I would call a "ratchet" view of evolution. He starts with Sewall Wright's picture of an adaptive landscape, with species sitting on the tops of peaks. Supposedly, one can slip somewhat down the side of such a peak, but never far enough that one would become extinct. Occasionally, one gets into a valley, and then shoots up the other side. There is no limit to the amount one can go up, and so taken over time, the total effect of evolution will be absolute rise. Yet while I would agree that in his major scientific work, *Variation and Evolution in Plants*, Stebbins keeps this thesis fairly well down -- apart from anything else, like all others he sees more progress in animals than plants -- it is certainly there. (See, for instance, the detailed discussion, from page 490 on, in *Variation and Evolution in Plants* as to how certain kinds of plants, generally recognized as "advanced," have managed to climb up certain steep "peaks," inaccessible to others.)

My lightly illustrated claim, therefore -- although hardly that light since I have been touching on absolutely seminal figures -- is that before 1966 evolutionary theory was (more cautiously, tended to be) a progressionist doctrine. We can turn now to today's evolutionists.

Progress Today

But, first, why choose 1966? I do so for the reason that this was the year of the publication of George Williams's *Adaptation and Natural Selection*. Deservedly, this book has the reputation of something which cuts through foggy or suspect biological thinking, like a hot knife through butter. After that book, you could never think of group selection with ease again. This is not to say that you could not be a group selectionist, but you had first to work through Williams's arguments and you may well have come out with a transformed position at the end.

Along with group selection, another of Williams's targets was biological progressionism. Up went the markers -- no less than five different versions of progressionism -- and all five were duly shot and knocked down. Or, at least, all five were shown to be lacking or unsubstantiated in various ways. Williams admitted candidly that his conclusions were not based on rigid deductions from the premises of evolutionary thought, nor were they based on definitive empirical evidence. Neither theory nor evidence is strong enough for this. Which point, incidentally, raises interesting and legitimate questions about Williams's own motivation. In fact, like T. H. Huxley (1901) before him, Williams is almost Manichean about the world around him, seeing it as innately horrible, if not evil. Our task is to combat nature, not to follow it docilely -- which sheep-like behavior is precisely what progressionists tend to urge upon us (Williams, in press).

But, whatever the exact connection between premises and conclusion, reading the pertinent chapter twenty years later, the effect is crushing, and was at that time acknowledged to be so. Progression is simply not on. So, what difference did Williams's critique make to evolutionary theorizing? Where do we stand now? To answer this question let me choose three high-profile biological contributors to *Molecules to Men*. This is surely a disinterested choice because the contributors were there precisely because they are at the top of the field of evolutionary studies. Moreover, let me make the task more difficult by not choosing someone like the late paleoanthropologist Glynn Isaac, on the fear that since he worked exclusively on humans he might thereby already be showing bias. (Misia Landau (1984) shows that this is almost an occupational disease of paleoanthropologists.) Also, I will take out Ernst Mayr on the grounds that his major scientific contributions were before 1966 (although, see Mayr 1984). I will even remove the geneticist Francisco J. Ayala, because he was a student and close associate of Dobzhansky. This last removal is a real blow, because Ayala is persistently and explicitly progressionist about the evolutionary process (e.g., Ayala 1982, this volume). What I will mention, however, is that as a coauthor of one of the major texts in the field, Ayala has done his part to put a favorable picture of progressionism before the eyes of the next generation of evolutionists (see Dobzhansky et al. 1977).

Start then with Edward O. Wilson, the entomologist and sociobiologist, and go to his major evolutionary work, *Sociobiology: The New Synthesis*, published in 1975. Everyone knows how controversial a book this was. Praised by some, it was excoriated by others, especially social scientists, humanists and radical biologists. What raised people's ire was Wilson's treatment of humans, for which he was called racist, sexist, reductionist, determinist, and various other -ists (Allen et al. 1977). But going back to the book, what strikes one -- at least, what strikes me -- is not the treatment of humans *per se*. After all, there is hardly an evolutionist who could keep his or her hands off the subject, and back in the early 1970s none of us were that ideologically pure. Rather, what is remarkable is the incredibly progressive nature of the book's treatment of evolution.

The very chapter titles and their order would have done credit to Lamarck: "the colonial microorganisms and invertebrates"; "the social insects"; "the cold-blooded vertebrates"; "the birds"; "evolutionary trends within the mammals"; "the ungulates and elephants"; "the carnivores"; "the nonhuman primates"; "man: from sociobiology to sociology." Moreover, the content does no less credit. There are, we learn, four "pinnacles" of social evolution, "high above the others" (Wilson 1975, p. 379). These are the colonial invertebrates, the social insects, the nonhuman mammals, and ourselves. But, this brings us face to face with a massive paradox.

> Although the sequence just given proceeds from unquestionably more primitive and older forms of life to more advanced and recent ones, the key properties of social existence, including cohesiveness, altruism, and cooperativeness, decline. It seems as though social evolution has slowed as the body plan of the individual organism became more elaborate. (Wilson 1975, p. 379)

What is the solution? In the case of the first three, the answer is relatively simple. In going from colonial invertebrates to mammals (although note how we are going from older "primitive" forms to newer "advanced" forms), we are going from the more related to the less. Slime molds and the like are collections of genetically identical entities. Thus, in helping others one is precisely helping oneself -- from a genetic point of view, that is. Moving to the social insects, particularly the hymenoptera, one is moving to organisms only partially overlapping genetically -- although as it happens in the case of sisters, the relationship is tighter than that which normally obtains. (This is the well-known 75% link, due to haplodiploidy.) So one gets sociality, but not as tight as in the invertebrate case. And then, finally, one has the nonhuman mammals and the even looser biological links which hold there -- and consequent looser sociality.

But -- and here progression gets right back on track -- humans have "been able to cross to [the] fourth pinnacle, reversing the downward trend of social evolution in general" (p. 382). How have we been able to do this, "the culminating mystery of all biology"? Obviously, this has come about through our increased brain power, our linguistic abilities, and all of the other things we associate with being distinctively human. Wilson proposes what he calls an "autocatalytic" process to account for human evolution. This is a kind of feed-back process (hypothesized with conscious reference to cybernetics), where once one is set on a particular track, evolution becomes self-reinforcing as success simply increases the need (and achievement) of success in the same direction.

Without going into details, which are hardly needed here, suffice it to say that Wilson sees two key points in our history. First, there was the transition from arboreal primates to real manlike apes (*Australopithecus*). In 1975, he dated this process from 15 through 5 million years ago, although like everyone else he would now bring the beginning date down significantly. This stage passed into a period of very rapid brain growth. Then, about 100,000 years ago, human evolution exploded forward again and we entered the cultural realm. Wilson speaks of "threshold effects" where the pressure builds up until you spurt forward in a new mode, and also of "multiplier effects" where fairly small scale genetic shifts can explode up causally into major cultural shifts (see also Wilson 1978).

As is well known -- almost too well known in that it draws attention from all else -- Wilson sees the end point of evolution as a fairly conventional, white, middle-class, Western, man-dominated state of affairs. Males have a propensity to sleep around, females to be coy. Males go out and hunt, whether this be in the Kalahari or on Madison Avenue, females stay at home with the children. Males especially have aggressive tendencies, although we do have a lot of built-in safety mechanisms. And -- quintessentially American -- we all have this burning urge to commit ourselves to God, or some such thing. Notoriously, we would rather believe than know. "Human beings are absurdly easy to indoctrinate -- they *seek it*" (Wilson 1975, p. 562, his italics).

Much more could be said, but just three quick points. First, interestingly, although our evolution is not yet over, there are hints that Wilson sees it ending in about a century. Either the process will end itself or outside forces (like the population explosion) will end it for us. ("But we have still another hundred years," p. 575.) Secondly, Wilson's progressionism fits smoothly with his world picture. Like J. Huxley, Wilson is an evolutionary humanist seeking for a secular alternative to Christianity (see especially Wilson 1978). Progress promises precisely this. Third, although Wilson leads and inspires this particular brand of sociobiology, not all beneath his shadow draw exactly the same conclusions. Sarah Hrdy, a primatologist, a feminist, and Wilson's student, thinks the coy, stay-at-home female is a chimera -- "the woman that never evolved" (Hrdy 1981). Apparently, because women conceal the point at which they ovulate, poor males are kept dancing -- very much at the mercy of female ends. Molecules to men transforms into worms to women.

Crossing the Atlantic, it is more difficult to make the progressionist case for my second evolutionist, Richard Dawkins. Indeed, in his recent book, *The Blind Watchmaker* (1986), he states flatly that "contrary to earlier prejudices, there is nothing inherently progressive about evolution" (p. 178). However, if the historians of science have taught us anything it is that things are not always quite what they seem, or rather how scientists describe their own work. For the moment, therefore, I shall pretend as if I were an historian in the year 2087 looking back on Dawkins's work. And the place I shall begin is with the fact that, although he denies progress in any absolute sense -- the monad-to-man sense -- Dawkins does agree that there is a kind of comparative progress in Darwin's special sense.

One would expect this concession given that Dawkins's position on adaptation is (like Darwin's) somewhere to the right of Thomas Aquinas. In particular, Dawkins explores favorably the notion of "arms races," where organisms compete (or "compete") against each other, thus giving rise to notions of direction and improvement as they strive (or "strive") to stay with or ahead of their opponents (or "opponents"). Lions and zebras, for example, or cheetahs and gazelles, compete against each other. The predators strive to get faster and in turn the prey also strives to move more quickly -- or, more accurately, since no one is thinking in Lamarckian terms, there is strong selective pressures on both predators and prey towards fleetness of foot (see also Dawkins and Krebs 1979.)

Dawkins (1986) reemphasizes that none of this adds necessarily to "a Victorian idea of the inexorability of progress" (p. 181), and in this he is surely right. An arms race might run off in all sorts of directions, many of which would be progressive in any absolute sense only to the most twisted of imaginations. Arms races can double back on themselves, undoing all that they have done. And they frequently stall or run right out of steam. Getting faster, for instance, puts all sorts of energy requirements on an organism. It might make a great deal of sense, biologically, to have and to raise more offspring, even if occasionally one goes to bed hungry or fails to go to bed at all.

And yet Darwin, as we saw, managed to turn his comparative progress towards absolute progress, and there are strains of this kind of thought in Dawkins also. On the one hand, there is the question of what humans are really good at, and the answer which comes back strongly is "brain power." A central place is given to Harry Jerison's (1985) notion of an "encephalization quotient" (EQ). This is a kind of measure of intelligence which abstracts the surplus available for sheer thinking over and above that necessary simply for getting the possessor's body to function. (Necessarily an elephant has a bigger brain than a mouse, simply because an elephant is so much bigger than a mouse. The question is how they compare when you have taken size into consideration.) As you might expect on this measure humans come out well ahead of the competition.

Measured EQs among modern mammals are very varied. Rats have an EQ of about 0.8, slightly below the average for all mammals. Squirrels are somewhat higher, about 1.5. Perhaps the three-dimensional world of trees demands extra computing power for controlling precision leaps, and even more for thinking about efficient paths through a maze of branches that may or may not connect farther on. Monkeys are well above average, and apes (especially ourselves) even higher. (Dawkins 1986, pp. 189-90)

Dawkins warns us against getting too carried away. That humans have an EQ of 7 and hippos of only 0.3 does not necessarily mean that we are over twenty times as bright as hippos. "But the EQ as measured is probably telling us *something* about how much 'computing power' an animal has in its head, over and above the irreducible minimum of computing power needed for the routine running of its large or small body" (p. 189).

On the other hand, having thus set the stage, Dawkins then goes on to give us a fairly clear idea of how he evaluates the end products of arms races. You might, for instance, prize thicker and better armor. Or you might value a disinclination to get into a fight. Or, then again, "computing power" might seem central to you -- as apparently it does to Dawkins. Picking up again on Jerison's work, this time on fossils, Dawkins notes how the arms race between carnivores and herbivores seems to lead to bigger and -- this is important -- better brains.

> [Jerison's] conclusion is, firstly, that there is a tendency for brains to become bigger as the millions of years go by. At any given time, the current herbivores tended to have smaller brains than the contemporary carnivores that preyed on them. But later herbivores tended to have larger brains than earlier herbivores, and later carnivores larger brains than earlier carnivores. We seem to be seeing, in the fossils, an arms race, or rather a series of restarting arms races, between carnivores and herbivores. This is a particularly pleasing parallel with human armament races, since the brain is the on-board computer used by both carnivores and herbivores, and electronics is probably the most rapidly advancing element in human weapons technology today. (Dawkins 1986, p. 190)

I really do start to get the sense that while all adaptations are equal, some are more equal than others. And to this I would add two related external items. First, Dawkins's use of Jerison's work is in itself hardly theory-neutral, given that such work is highly controversial and has been much criticized as far too simplistic. Let me rush to say that Jerison may be right and in any case Dawkins does acknowledge the controversy; but, as things stand, given that Dawkins does use Jerison as centrally as he does, I sense someone who has a special case to make. And this ties in with my second item, namely that the thesis that Dawkins prizes computing abilities above other abilities certainly fits in with his general intellectual approach to life. Apart from a longtime enthusiasm for computers, *The Blind Watchmaker* itself tells us this much. The genius of the book lies in the third chapter, where Dawkins shows how the home computer is a metaphor for our time, and how through such a metaphor one can give the lie to so much silliness which is spoken about the impossibly improbabilistic nature of Darwinism.

All in all, I would have to agree with our historian of 2087 if he or she were to conclude that Dawkins is not so very far from the grand tradition of Darwinian progressionists. And this brings me, as we cross back over the Atlantic, to the third, final, and most difficult case of all: Stephen Jay Gould. Again we have someone who denies he is a progressionist. Indeed, he describes the notion as a "noxious, culturally embedded, untestable, nonoperational, intractable idea that must be replaced if we wish to understand the patterns of history" (Gould, this volume). Again, however, I would argue that there is more to the case than meets the eye and that, in respects, Gould is very much a progressionist. Certainly he is not a progressionist in the Wilsonian sense. Anything but, for he has been one of sociobiology's most persistent critics. Yet, the label applies, nonetheless. In addition to the points about to be made, Dawkins and Krebs (1979) point out that Gould (1977a) contains discussion of predator-prey arms races.

As is well known, Gould has been the originator and skillful promoter of the so-called theory of "punctuated equilibria," a view of the course of evolution (as revealed through the fossil record) as one of abrupt jerkiness, of rapid change linking long periods of inactivity ("stasis"). It is a theory opposed to "phyletic gradualism," which sees change as ongoing and continuous. Even in the theory's short history (less than two decades), I see significant development, going through some three phases (Ruse, in press). First, the theory (as conceived by Gould and his cotheorist Niles Eldredge) was considered to be a fairly direct extension of the synthetic theory, with Mayr's founder principle playing a crucial role -- it led to rapid speciation, the only point at which significant evolutionary change occurred (Eldredge and Gould 1972). Second, towards the end of the 1970s, Gould began to emphasize the nonadaptive, discontinuous nature of his theory. I do not think Gould ever became an anti-Darwinian or an outright saltationist (of the Richard Goldschmidt kind), but certainly he began to think of the evolutionary process through such lenses (e.g., Gould 1979, 1980a). Then, rapidly, at the beginning of this decade, came the synthesis, with punctuated equilibria theory being seen as an "expansion" of Darwinism -- picking up on adaptive evolution through natural selection but going beyond (Gould 1982b).

What direction does such expansion take? Most crucially, that we must now think of evolutionary processes in a *hierarchical* fashion. The organic world is layered -- molecules, genes, cells, organisms, populations, species, and more. What we must see is that although processes at one level are clearly dependent on processes at lower levels, they cannot be reduced to such levels. Things happening at higher levels are in some sense autonomous. In particular, while it is undoubtedly true that we have selection working in general and/or on organisms producing adaptations, there are processes which may or may not promote adaptation at higher levels. And, one such level is clearly the human cultural domain.

> No one doubts, for example, that the human brain became large for a set of complex reasons related to selection. But having reached its unprecedented bulk, it could, as a computer of some sophistication, perform in an unimagined range of ways bearing no relation to the selective reasons for initial enlargement. Most of human society may rest on these non-adaptive consequences. How many human institutions, for example, owe their shape to that most terrible datum that intelligence permitted us to grasp -- the fact of our personal mortality (Gould 1982b, p. 384).

There is a great deal more in this vein, particularly when Gould is less concerned with pushing his own positive view of evolution (as he is in the just-quoted passage) and more concerned with reacting negatively against human sociobiology. But, the overall message is the same: "I welcome the coming failure of reductionistic hopes because it will lead us to recognize human complexity at its proper level" (Gould 1980b, p. 266).

I argue that this vision of evolution is progressive. It may not be progressive in a traditional monad-to-man fashion. Indeed, it must be acknowledged that recently Gould has been arguing that, if anything, evolution is bottom heavy, in that new types seem to flourish most in their early stages rather than their later stages (Gould et al. 1987, this volume). (That such conclusions strike me as artifactual is beside the point.) Nevertheless, apart from noting with some interest the significance of the term "complexity" in the last quote -- complexity is often that towards which progress is supposed to lead -- the very idea of a hierarchy is something which incorporates the notion of an advance. The best-known hierarchy outside biology is the Catholic Church, and if the Pope is not above the common parish priest I do not know what is.

Of course, a hierarchy might just involve a conceptual advance and not a temporal one -- the first Pope supposedly was Peter and he came before all the parish priests. I suspect that Gould could complain that his hierarchy is likewise more conceptual than temporal. But the fact is that we humans do seem to sit somewhere up the chain -- sociobiology apparently will work with nonhumans even if it does not apply to us. (Do note, incidentally, how Gould in one of the quotes just given above echoes Dawkins in picking up on the computer-like nature of our brains, obviously thinking this point significant.)

And, in any case, you cannot avoid the temporal dimension entirely. Selfish genes preceded selfish humans. As John Maynard Smith (this volume) has said, in a discussion essentially favorable to the notion of progress as a hierarchical phenomenon: "Although biologists may be reluctant to see [the stages through which life has passed] as representing 'advance,' they are progressive in one sense: the sequence in which they have occurred is not arbitrary, since each stage was a necessary precondition for the next" (p. 219).

One final point. With my other evolutionists, I have tried to show how their scientific views mesh with their general philosophies. Gould has made comments favorable to Marxism, so you might suspect that I suggest that truly he sees life ending with us all hunting in the morning, fishing in the afternoon, tending cattle in the evening and doing philosophy after supper. However, while I think it legitimate to note the temporally directed thinking of such staunch antireductionists as Marx and Engels (Graham 1972), I suspect it is more profitable to see both them and Gould as part of a general European view of life, a view which goes back to such thinkers as Hegel and the earlier nineteenth-century *Naturphilosophen*. Certainly, by his own admission, this is where Gould's vision of evolution fits -- and here we do find a notion of advance which merits the name "progress" (Gould 1983). (In this context, think of Louis Agassiz 1859; see also Bowler 1976.)

The Persistence of Progress

I claim that evolutionary thought has, through its history, been permeated by thoughts of progress, and that this still holds true today. I certainly do not claim that every (scientific) evolutionist is a progressionist, but many are -- and this includes important thinkers. I recognize that today biologists tend to downplay their progressionism, not wanting to seem stuck in the nineteenth century or at one with religious visionaries like Teilhard. There are outright denials, or claims that thoughts about progression are not part of science, or (as we have seen) the shunting of such thoughts to "popular" books.

However, as I have intimated (although hardly proven in any definitive fashion) the separation is nothing like as absolute as all of this. I argue that, taken as a whole, the picture is one of progression, and I would go on to say that the "real" science is not all that value-free. Of course, the average scientist working away on his or her fruitflies is rarely obsessed with immediate hopes of utopia. Nor do progressivist yearnings get on every page of *Evolution*. But they are there. To take but one example, Dawkins started expressing his views on progress in the pages of the *Proceedings of the Royal Society*. If that is not a home of real science I do not know what is.

I have shown (although again hardly proven in any definitive fashion) that hopes of progress are certainly not there on the surface of evolutionary phenomena, waiting to be read straight off. I take it there is now little doubt that, in major part at least, biologists' thoughts on evolutionary progress reflect or echo their views on progress generally -- and that such general views tend not to be entirely unsympathetic to the notion. As a philosopher, speaking formally, I do not find the situation in evolutionary biology at all surprising. Thanks to the work of people like Pierre Duhem (1914) and Willard van Orman Quine (1953), we are now aware fully that scientific theories are "underdetermined" by the facts -- that science has to outstretch its empirical base in order to achieve its goals. Moreover, values enter into science at this point of stretching, as one constructs the theoretical bricks from too little factual straw. It is now realized even by philosophers that the values, as in our case, will sometimes go beyond dry "epistemic" values like appeals to simplicity and predictive fertility and so forth. Science reflects all of those concerns that scientists like other humans, hold dear (Graham 1981).

Philosophically, what is perhaps surprising is the way that such a value-laden notion as progress has persisted in science. There have been claims that "mature" science strips away all value constraints except those such as simplicity (McMullin 1983). It may, of course, be the case that evolutionary biology uniquely suffers from a case of arrested development in this respect. Yet, although I have not addressed such issues directly, I (for one) would like to see some substantive and documented evidence about other areas of science before any final conclusions are drawn. I have made a case for biology. Now let others make a case for (say) physics. All I want to reiterate is what I said at the beginning. My aim in this discussion has been towards understanding, not criticism. I want to show you that there are thoughts of progress, not that these are necessarily wrong. Every copy of *The Blind Watchmaker* carries on its cover strong praise by me.

In any case, my primary concern here is not with philosophy but with biology. My main question is why the notion of progress persists. Why are evolutionary biologists such ardent progressionists? I know or suspect the answer in individual cases. Wilson, I have said, is seeking a secular religion -- one which will replace the fundamentalist Baptist belief of his Alabama youth. I even suspect that his hints about progress shortly ending are predicated on a hazy dispensationalist foreboding of a sort of secular Armageddon. It is for this religion-substitute-seeking reason that those who condemn Wilson for being a neo-conservative ideologue completely miss the mark (Segerstrale 1986). What I want to know now is why evolutionists as a breed seem to be such progressionists.

In the midnineteenth century, such a question would have made little sense. Everybody was! But, as we draw towards the end of the twentieth century, in the shadow of the Bomb, and overpopulation, and pollution, and AIDS, the question becomes pressing. Things are not as uniformly black as they were during the Wars or the Depression -- although remember that even then Fisher thought there was hope. Today, there is a spectrum. Some people are optimists. For instance, Third World leaders and thinkers are almost universally progressionists (Young 1982). In a way, they have to be. Yet many others, especially in the West, surely echo the sensitive Jewish-Canadian philosopher Emil Fackenheim (1968).

> A view still popular in America holds that history progresses necessarily but intermittently; relapses may occur, but these become ever less serious. But to me Nazism was, and still is, not a relapse less serious than previous relapses, but a total blackout . . . . History is regarded as necessary progress only by those who are relatively remote from the evils of history . . . . The real conclusion I derived . . . is that if even a single brave and honest deed is in vain, if a single soul's unjust suffering goes unredeemed, that then history as a whole is meaningless. With this conclusion, the progress view of history, so far as I was concerned, had suffered total shipwreck. (p. 85)

I ask again: Why are evolutionists progressionists? Why do they fall into the positive camp rather than the negative one? I take it that now we start to look for reasons within or connected to evolutionary theorizing itself. It seems there is something about being an evolutionist which inclines you to progressionism, which of course may well infect your general thinking and rebound back. But, unlike the nineteenth century we cannot just look to the outside for initial thoughts of progress. In a tentative way, I make three suggestions.

First, I think we have at work here something akin to (or perhaps a variant of) what cosmologists and natural theologians these days refer to as the "Anthropic Principle" (Barrow and Tipler 1986). This comes in many versions (too many, for my taste), but it is agreed that the Principle is something which sets conditions or constraints on the very possibility of knowledge: "What must the world be like in order that man may know it"? (p. 143). I suggest that in the case of evolution the problem is that we humans are not disinterested observers from afar. We are products of the process ourselves and, although fortunately for our well-being, unfortunately for our understanding we are necessarily end products, in that we are still around to ask questions. Also, we are necessarily good (or if you like, "good") products, in that we are able to ask questions.

It is not necessary that we think ourselves the very best products. If there were angels around, presumably we would think ourselves second-class. (In medieval times, humans did only come halfway up the chain of being, and today Caucasians are having to grapple with the prospect that possibly

Orientals are innately more intelligent.) But the point is that, by the very nature of the case, just as (in Cartesian fashion) merely asking if you exist pushes you towards a positive answer, likewise, merely asking if there has been progress (and if we figure in it) pushes you towards a positive answer. As I once naively asked elsewhere, not realizing then that this probably tells you more about me than about objective reality: if the brutes disagree, then why don't they say so? (Ruse 1985).

Second, and here obviously I speak more to those with Darwinian inclinations, it is clear that the whole notion of progress as comparative success comes right out of the mechanism of natural selection. This mechanism speaks uniquely and directly to one thing: adaptation. It was this which Darwin intended and prided himself on, and it is this which today's ultra-Darwinians like Richard Dawkins pick up and push. But what is adaptation all about? What is the essence of an adaptive characteristic? Simply that we have before us something which is design-like (Williams 1966; Ruse 1982). Paradigmatic adaptations are as if they were created by intention, as if they were artifacts. Indeed, before Darwin they were thought to be artifacts -- God's artifacts. This is why Dawkins (1983) can refer to himself jocularly but truly as a transformed Paleyist. Like the transformed cladists' attitudes towards taxonomy, he recognizes the phenomena even though he wants to give a radically different interpretation from that of a Christian teleologist like Archdeacon Paley.

But seizing on the artifact-metaphor at the heart of Darwinism points us to two things. On the one hand, we see that we move beyond the evidence. Whether one could ever have an adequate theory of evolution entirely without the metaphor is perhaps a moot point, although some like Gould think it much overdone. But, this apart, it is indubitable that inasmuch as one construes organisms as if they were designed, one is interpreting. And, with this point established, on the other hand we see that such interpretation does push us towards progress of a kind. It is of the very nature of artifacts that we humans try to improve them. This might be done dramatically through an arms race, as Dawkins suggests; although, I am not sure that such a race is always necessary. However, whatever the proximate causes, artifacts do get improved -- they have a progressive history -- and this I suggest tips evolutionists, Darwinians especially, into progressionism.

Yet, there is artifactual progress and artifactual progress. To get from A to B, you might go by bicycle. You might go by Concorde. All of us would say that bicycles are sometimes superior. Some of us would say that bicycles are always superior. Some of us would say that, for all of the virtues of bicycles, Concorde represents a real advance. Those Darwinians who move from comparative success to absolute progress seem to fall into this last group. The question I ask is, why is it that such Darwinians seem to be most or nearly all Darwinians? And, this translates into the related question as to why it is that high-powered technology has such a hold over evolutionists? (Or why it has a hold over so many evolutionists, for I do not want to make absolute claims.)

My surmise, and here I make my third and final suggestion, is simply that science in general -- of which evolutionary theory is a case in particular -- is the one area of human endeavor where we do seem to get unambiguous progress (Maynard Smith 1982; Edsall 1982; Ruse 1986). The discovery of the DNA model, for instance, is a real advance over premolecular genetics, as was Einsteinian relativity over Newtonian mechanics. In Darwinian evolutionists, therefore, we have people who know success, appreciate success, and expect success. It is consequently natural that they should be attracted to sophisticated high-powered technology, for it is precisely this which rides on the back of pure-science advance, apart from being that which is often necessary for further scientific advance. Hence, Darwinian evolutionists -- like the good scientists that they are -- read the progress of their science out into everything else, and through technology into the subject of their studies. Perhaps the case is broader then this, even. You do not have to be an ultra-Darwinian to appreciate the success of science. It seems plausible, therefore, that short-cutting the detour through technology, we have here a reason why all evolutionists are attracted to progress. What works in the world of thought, must surely work in the world of organisms.

Lest you think I exaggerate let me draw you back to the late Sir Peter Medawar, quoted at the beginning of this paper. He was no foggy-minded mystic. Indeed, he had the most sneering things to say of Huxley's infatuation with Teilhard (Medawar 1969). Yet he believed in progress, ultimately seeing the biological in a fusion with the cultural. And, significantly, this paean to progression came in his presidential address to the British Association, a point at which (by his own admission) he spoke "*of* and *for* science" (Medawar 1972, p. 9, his italics). I rest my case.

References

Agassiz, L. 1859. *Essay on classification*. London: Longman, Brown, Green, Longmans, and Roberts and Trübner.

Allen, L., et al. 1977. Sociobiology: A new biological determinism. In *Biology as a social weapon*, ed. Sociobiology Study Group of Boston. Minneapolis: Burgess.

Almond, G., M. Chodorow and R. H. Pearce. 1982. *Progress and its discontents*. Berkeley: University of California Press.

Ayala, F. J. 1982. The evolutionary concept of progress. In *Progress and its discontents*, ed. G. Almond, M. Chodorow and R. H. Pearce, 106-24. Berkeley: University of California Press.

Barrow, J., and F. Tipler. 1986. *The anthropic cosmological principle*. Oxford: Oxford University Press.

Bendall, D. 1983. *Evolution from molecules to men*. Cambridge: Cambridge University Press.

Bennett, J. H. 1983. *Natural selection, heredity, and eugenics*. Oxford: Oxford University Press.

Bowler, P. J. 1976. *Fossils and progress*. New York: Science History Publications.

Bowler, P. J. 1984. *Evolution: The history of an idea*. Berkeley: University of California Press.

Cannon, W. F. 1961. The impact of uniformitarianism. Two letters from John Herschel to Charles Lyell, 1836-1837. *Proceedings of the American Philosophical Society* 105:301-14.

Chambers, R. 1844. *Vestiges of the natural history of creation*. London: Churchill.

Chambers, W. 1872. *Memoir of Robert Chambers, with autobiographic reminiscences of William Chambers LLD*. Edinburgh: Chambers.

Darwin, C. 1859. *On the origin of species by means of natural selection*. London: John Murray.

Darwin, C. 1871. *Descent of man*. London: John Murray.

Darwin, C. 1959. *The origin of species by Charles Darwin: A variorum text*, ed. M. Peckham. Philadelphia: University of Pennsylvania Press.

Darwin, C. 1960. Darwin's notebooks on transmutation of species. Part I. Notebook 'B', ed. G. de Beer. *Bulletin of the British Museum, Natural History Series* 2:27-73.

Darwin, C., and A. R. Wallace. [1858] 1958. *Evolution by natural selection*. Facsimile? Cambridge: Cambridge University Press.

Dawkins, R. 1983. Universal Darwinism. In *Evolution from molecules to men*, ed. D. S. Bendall. Cambridge: Cambridge University Press.

Dawkins, R. 1986. *The blind watchmaker*. London: Longmans.

Dawkins, R., and J. R. Krebs. 1979. Arms races between and within species. *Proceedings of the Royal Society of London*, B, 205:489-511.

Dobzhansky, T. 1962. *Mankind evolving*. New York: Columbia University Press.

Dobzhansky, T. 1967. *The biology of ultimate concern*. New York: New American Library.

Dobzhansky, T., F. J. Ayala, G. L. Stebbins, and J. W. Valentine. 1977. *Evolution*. San Francisco: W. H. Freeman.

Duhem, P. 1914. *The aim and structure of physical theory*. Princeton: Princeton University Press.

Edsall, J. 1982. Progress in our understanding of biology. In *Progress and its discontents*, ed. G. Almond, M. Chodorow and R. H. Pearce, 135-60. Berkeley: University of California Press.

Eldredge, N., and S. J. Gould. 1972. Punctuated equilibria: An alternative to phyletic gradualism. In *Models in paleobiology*, ed. T. J. M. Schopf. San Francisco: Freeman Cooper.

Fackenheim, E. L. 1968. *Quest for past and future: Essays in Jewish theology*. Boston: Beacon.

Fisher, R. A. 1930. *The genetical theory of natural selection*. Revised and reprinted 1958. New York: Dover.

Fisher Box, J. 1978. *R. A. Fisher. The life of a scientist*. New York: Wiley.

Gould, S. J. 1977a. *Ever since Darwin*. New York: Norton.

Gould, S. J. 1977b. *Ontogeny and phylogeny*. Cambridge: Harvard University Press.

Gould, S. J. 1979. Episodic change versus gradualist dogma. *Science and Nature* 2:5-12.

Gould, S. J. 1980a. Is a new and general theory of evolution emerging? *Paleobiology* 6:119-30.

Gould, S. J. 1980b. Sociobiology and the theory of natural selection. In *Sociobiology: Beyond nature/nurture?* ed. G. Barlow and J. Silverberg, 257-69. Boulder, CO: Westview.

Gould, S. J. 1982a. *The mismeasure of man*. New York: Norton.

Gould, S. J. 1982b. Darwinism and the expansion of evolutionary theory. *Science* 216:380-7.

Gould, S. J. 1983. The hardening of the modern synthesis. In *Dimensions of Darwinism*, ed. M. Grene, 71-93. Cambridge: Cambridge University Press.

Gould, S. J., N. L. Gilinsky, and R. Z. German. 1987. Asymmetry of lineages and the direction of evolutionary time. *Science* 236:1437-41.

Graham, L. R. 1972. *Science and philosophy in the Soviet Union*. New York: Knopf.

Graham, L. R. 1981. *Between science and values*. New York: Columbia University Press.

Greene, J. 1981. *Science, ideology and world view*. Berkeley: University of California Press.

Haeckel, E. 1883. *History of creation*. 3d ed. London: Kegan Paul, Trench.

Houghton, W. E. 1957. *The Victorian frame of mind*. New Haven: Yale.

Hrdy, S. 1981. *The woman that never evolved*. Cambridge: Harvard University Press.

Huxley, J. S. 1942. *Evolution: The modern synthesis*. London: Allen and Unwin.

Huxley, T. H. 1901. *Evolution and ethics, and other essays*. London: Macmillan.

Hyatt, A. 1889. Genesis of the Arietidae. *Bulletin of the Museum of Comparative Zoology* 16(3): 1-238.

Jerison, H. J. 1985. Issues in brain evolution. In *Oxford surveys in evolutionary biology*, ed. R. Dawkins and M. Ridley. 2:102-34.

Jordanova, L. 1984. *Lamarck*. Oxford: Oxford University Press.

Lamarck, J.-B. 1809. *Zoological philosophy*. Trans. H. Elliot, 1963. New York: Hafner.

Landau, M. 1984. Human evolution as narrative. *American Scientist* 72 (3):262-8.

Lovejoy, A. O., and G. Boas. 1935. *Primitivism and related ideas in antiquity*. Baltimore: Johns Hopkins University Press.

Lyell, C. 1830-33. *Principles of geology*. London: John Murray.

Maynard Smith, J. 1982. Storming the fortress. *The New York Review of Books*. May 13:41-2.

Mayr, E. 1942. *Systematics and the origin of species*. New York: Columbia University Press.

Mayr, E. 1963. *Animal species and evolution*. Cambridge: Harvard University Press.

Mayr, E. 1982. *The growth of biological thought*. Cambridge: Harvard University Press.

Mayr, E. 1984. Evolution and ethics. In *Darwin, Marx and Freud*, ed. A. Caplan and B. Jennings, 35-46. New York: Plenum.

McMullin, E. 1983. Values in science. In *PSA 1982*, ed. P. Asquith and T. Nickles. East Lansing, MI: Philosophy of Science Association 2:3-28.

Medawar, P. 1969. *The art of the soluble*. Harmondsworth, Middlesex: Penguin.

Medawar, P. 1972. *The hope of progress*. London: Methuen.

Osborn, H. F. 1929. *From the Greeks to Darwin*. New York: Scribner's.

Quine, W. 1953. Two dogmas of empiricism. In *From a logical point of view*, ed. W. Quine. Cambridge: Harvard University Press.

Rudwick, M. J. S. 1969. The strategy of Lyell's *Principles of geology*. *Isis* 61:5-33.

Ruse, M. 1979. *The Darwinian revolution: Science red in tooth and claw*. Chicago: University of Chicago Press.

Ruse, M. 1982. Teleology redux. In *Scientific philosophy today*, ed. J. Agassi and R. S. Cohen, 299-309. Dordrecht: Reidel.

Ruse, M. 1985. Is rape wrong in Andromeda? Reflections on extra-terrestrial life. In *The search for extra-terrestrial life*, ed. E. Regis. New York: Cambridge University Press.

Ruse, M. 1986. *Taking Darwin seriously: A naturalistic approach to philosophy*. Oxford: Blackwell.

Ruse, M. In press. Is the theory of punctuated equilibria a new paradigm? *Journal of Social and Biological Structures*.

Schweber, S. 1977. The origin of the *Origin* revisited. *Journal of the History of Biology*. 10:229-316.

Schweber, S. 1980. Darwin and the political economists: Divergence of character. *Journal of the History of Biology* 13:195-289.

Segerstrale, U. 1986. Colleagues in conflict: An 'in vivo' analysis of the sociobiology controversy. *Biology and Philosophy* 1:53-88.

Simpson, G. G. 1944. *Tempo and mode in evolution*. New York: Columbia University Press.

Simpson, G. G. 1953. *The major features of evolution*. New York: Columbia University Press.

Simpson, G. G. 1963. *This view of life*. New York: Harcourt, Brace and World.

Spencer, H. 1852. A theory of population, deduced from the general law of animal fertility. *Westminster Review*, n.s., 1:468-501.

Spencer, H. [1857] 1868. Progress: Its law and cause. *Westminster Review*. Reprinted in *Essays: Scientific, political and speculative*, 1:1-60. London: Williams and Norgate.

Stebbins, G. L. 1950. *Variation and evolution in plants*. New York: Columbia University Press.

Stebbins, G. L. 1969. *The basis of progressive evolution*. Charleston, NC: University of North Carolina Press.

Teilhard de Chardin, P. 1959. *The phenomenon of man*. London: Collins.

Wassersug, R., and M. Rose. 1984. A reader's guide and retrospective to the 1982 Darwin centennial. *Quarterly Review of Biology* 59:417-37.

Williams, G. C. 1966. *Adaptation and natural selection*. Princeton: Princeton University Press.

Williams, G. C. In press. Huxley's evolution and ethics in sociobiological perspective. *Zygon* forthcoming.

Wilson, E. O. 1975. *Sociobiology: The new synthesis*. Cambridge: Harvard University Press.

Wilson, E. O. 1978. *On human nature*. Cambridge: Harvard University Press.

Young, R. M. 1969. Malthus and the evolutionists: The common context of biological and social theory. *Past and present* 43:109-45.

Young, C. 1982. Ideas of progress in the Third World. In *Progress and its discontents*, ed. G. Almond, M. Chodorow, and R. H. Pearce. Berkeley: University of California Press.

HISTORICAL AND
COMPARATIVE STUDIES

The Moral Foundations of the Idea of Evolutionary Progress: Darwin, Spencer, and the Neo-Darwinians

Robert J. Richards

An eminent evolutionary thinker has interpreted natural selection as the muscle producing biological and social progress. He claims that "as natural selection works solely by and for the good of each being, all corporeal and mental endowments will tend to progress towards perfection." Most historians, philosophers, and biologists, however, would regard attaching the idea of progress to Darwin's theory as comparable to stitching a Victorian bustle on the nylon running shorts of a woman marathoner, a cultural atavism disguising the slim grace supplying the real power. Progress becomes Herbert Spencer's evolutionary theory, but that is a museum piece long ago shelved. Darwin's still vital conception, by contrast, threatens notions of progress embedded in mid-Victorian culture. "Darwin's mechanism," Peter Bowler has recently insisted (1986, p. 41), "challenged the most fundamental values of the Victorian era, by making natural development an essentially haphazard and undirectional process." This historical assessment receives support from probably the most influential logical analysis of Darwinian theory, that of George Williams.

In *Adaptation and Natural Selection* (1966, pp. 34-55), Williams isolates several simple features of the evolutionary process that would seem to defeat any possibility of long-term progressive development. First is the structure of the mechanism itself: natural selection is the substitution of alleles more favorable in a given environment for those less favorable. Since the selective environment constantly changes, what was adaptive in the past must become increasingly less so. Add to this the thermodynamical wheeze of entropy (see Wiley, this volume) -- indicated by genetic mutation and recombination -- well, the situation would appear as Leigh Van Valen (1973) has characterized it.

129

The Red Queen has to keep running on the slipping treadmill of environmental change just to remain in the same place -- no long-term progress appears possible. In addition to his a priori analysis of the possibility of evolutionary progress, Williams examines three candidates for empirically grounding the idea: progress as accumulation of genetic information, progress as morphological complexity, and progress as effectiveness of adaptation. All of these, at one time or another, have been proposed as measures of evolutionary advance. But these three candidates succumb to those structural features of the evolutionary process just mentioned. Further, Williams has little trouble in discovering obvious counter-examples to each. For instance, human beings may be more advanced than apes in their effective adaptation to a broad range of environments -- so you might mark the transition from *Australopithecus afarensis* to *Homo sapiens* a progressive one. But where you find man you also find the cockroach. And the cockroach will dwell happily in environments quite lethal to our species. According to the standard of effective adaptation, cockroaches may be the most important product of progress.

Let me mention one more recent examination of the idea of evolutionary progress -- this by one who is, like Bowler, a historian, but also like Williams, a scientist. Ernst Mayr, in his *Growth of Biological Thought* (1982), has considered both the historical and the scientific questions, namely what was Darwin's view of progress and what should we make of the idea of evolutionary progress. Not surprisingly, Mayr thinks we should believe what Darwin did. He maintains that "Darwin, fully aware of the unpredictable and opportunistic aspects of evolution, merely denied the existence of a lawlike progression from 'less perfect to more perfect'" (1982, p. 531). Darwin strongly objected to any notions of "an intrinsic drive to perfection, controlled by 'natural' laws," says Mayr (p. 532). Darwin might, as a cultural afterthought, have occasionally referred to evolutionary progress, but such short-term evolutionary innovations that every biologist must recognize were only the "a posteriori results of variation and natural selection" (p. 532). Evolutionary laws leading to biological progress have been promulgated by Spencer and some minor Darwinians, but, it is believed, the master steered clear of that slope which leads to a yawning abyss of Social Darwinism and Sociobiology (but see Ospovat, 1981, pp. 210-28, for a different analysis of Darwin's notions about progress).

The quotation with which I began -- that "as natural selection works solely by and for the good of each being, all corporeal and mental endowments will tend to progress towards perfection" -- of course, comes from the *Origin of Species* (1859, p. 489). Perhaps one might simply chalk it up to some carelessness on the part of Darwin, maybe even a sop to the theologically blinkered. But I don't think so. The delicate scroll work and the emblems depicting cultural advance etched into the machinery of natural selection were not merely decorative motifs; they depicted functional parts of his device. Indeed, Darwin crafted natural selection as an instrument to manufacture biological progress and moral perfection. In this respect, his theory does not substantially differ from Spencer's, upon which much abuse is often heaped for making evolution necessarily progressive. Let me begin, then, by briefly considering the ways in which a moral vision structured Spencer's own theory of evolution. His case is instructive for understanding the historical logic of Darwin's theory.

The Moral Foundations of Spencer's Theory of Evolution

Spencer's evolutionary theory is generally thought to have nasty moral and social consequences. Typically he is understood to have formed his moral conception around a razor-edged theory of survival of the fittest, an embrace most people think could only have deadly consequences for social philosophy. I believe, on the contrary, that the relation between Spencer's evolutionary theory and his moral theory is just the reverse of what is usually supposed. Spencer first formed a utopian social philosophy that would bear striking resemblance to that of his two contemporaries working at the British Museum, two Germans plotting revolution for their fatherland, Marx and Engels. Unlike their conception, though, Spencer's social theory was braced by distinctively English theological and moral principles. It was on this theological and moral loom that the fabric of his evolutionary theory was woven (Richards 1987, chaps. 6 and 7; Peel 1971).

Spencer grew up in the English Midlands, the country of Erasmus Darwin, James Watt, and Josiah Wedgwood -- those practical-minded men who founded the Derby Philosophical Society, which had Spencer's father as its recording secretary. The Spencer family counted itself among John Wesley's earliest followers. The father, though, dissented even from such heterodox association, and thereby set the son on the nonconformist's path. In his youth, Spencer became involved in several nonconformist agitation groups, often with his uncle Thomas Spencer serving as leader. It was his uncle, a curate with a parish near Bath, who arranged for the twenty-one-year-old to contribute a series of letters to the newspaper *The Nonconformist* in 1842. These letters, bearing the title "The Proper Sphere of Government," set down the theological, moral, and social considerations that served as templates for later developments of Spencer's theory of evolution.

In the letters (1842, 15 June to 23 November), Spencer argued that natural laws divinely designed for man's happiness, which regulated the physical, organic, and mental realms, governed the social realm as well. If God had established the laws of these dominions, only mischief would arise when human beings attempted to tinker with them. Government interference in the social sphere, Spencer believed, deformed the self-correcting natural forces controlling social development. If natural forces were left to play out their roles, then one could expect a divinely ordained consummation. Exactly what that terminus of social development might consist of, Spencer specified more precisely nine years later, in 1851, in his book *Social Statics*. Progressive development, both in the organic and the social realms, would lead to a classless society in which each person would freely exercise his or her talents, limited only by the freedom of others; it would be a society in which the state had withered away, land would be held in common, and women and children would have their freedom equally respected. This Godly conclusion and the natural laws leading to it demonstrated, according to Spencer, that:

> Progress . . . is not an accident, but a necessity. Instead of civilization being artificial, it is a part of nature; all of a piece with the development of the embryo or the unfolding of a flower. The modifications mankind have undergone, and are still undergoing, result from a law underlying the whole organic creation; and provided the human race continues, and the constitution of things remains the same, those modifications must end in completeness . . . . So surely must the human faculties be molded into complete fitness for the social state; so surely must the things we call evil and immorality disappear; so surely must man become perfect. (1851, p. 65)

Early in the 1840s, Spencer had read Lyell's (1830-1833) description and refutation of Lamarckian transformation theory. In 1844, he enjoyed the speculative whimsy of Robert Chambers, whose *Vestiges of the Natural History of Creation* (1844) set forth an evolutionary conception of organic life -- a conception that brought quick denunciation from British zoologists and gloomy regret from Charles Darwin, who feared Chamber's book might sink in ridicule his own evolutionary craft. Though Spencer's *Social Statics* resonated finely to these earlier evolutionary proposals, no one immediately felt the vibrations. Those few critics who took notice of the book were hardly disturbed by its faint evolutionary tremors, but they were quite agitated by the martial tempo of its anarchistic socialism. The author appeared as a British Proudhon holding up a moderately bloody fist.

As the result of attending lectures by Richard Owen, Britain's leading morphologist and student of Georges Cuvier, and of reading in physiology -- especially the books of William Carpenter (1841) and Henri Milne-Edwards (1841) -- Spencer began consciously and more expressly to formulate his evolutionary ideas. In a series of essays in the early 1850s, in his *Principles of Psychology* (1855), and in his metaphysical work *First Principles* (1862), he laid down those evolutionary conceptions that led Alexander Bain to call him "the philosopher of the Doctrine of Development, notwithstanding that Darwin has supplied a most important link in the chain" (1863, MS. 791, no. 67).

The biosocial evolutionary theory that Spencer began to elaborate in the early 1850s was powered by a law that would produce for Spencer -- and, I believe, for Darwin -- the kind of evolutionary progress that their theological and ethical sentiments demanded. Spencer understood the law to govern "an advance from homogeneity of structure to heterogeneity of structure"; and he believed that this "law of organic progress is the law of all progress" (1857, p. 446). He derived his principle in part from von Baer's and Milne-Edwards's conception of a division of labor in organic structures. Spencer argued that continued adaptation of organisms to complex environments -- chiefly through functional acquisitions of heritable characteristics, the Lamarckian device -- would produce structures more adapted to deal successfully with those environments. It would create both organisms and societies of organisms that displayed greater specialization of parts within an overall integration of functions -- more complex organisms and societies, that is, more progressive and perfect organisms and societies. As Spencer concluded his article *Progress: its Law and Cause*, "progress is not an accident, not a thing within human control, but a beneficent necessity" (1857, p. 484).

Spencer is usually supposed to have formulated his ethical ideas in light of evolution by natural selection, so that the leading principle of his moral philosophy is assumed to be the survival of the fittest. Even his friends, Thomas Huxley for instance, later read Spencer as advocating a brutal individualism according to which survival would go to those most fit for the Hobbesian struggle, instead of to those who were most morally fit (Huxley 1902, pp. 46-116). But this reading really does misrepresent Spencer's position. He supposed, except in one particular instance, that the leading principle accommodating man to harmonious social development was not competitive struggle but Lamarckian absorption of the principles of justice and fair treatment. The one conspicuous use Spencer made of Darwin's device of natural selection was in the form of community selection (ultimately group selection), by which the sentiments of altruism, that pivot of the moral life, might be established, so as to direct a society to ends dictated by the ideals of justice and greatest freedom. The passage to this end would be guaranteed by the unencumbered evolutionary process, which, as Spencer envisioned it in the last sentence of his *Principles of Psychology*, is a "grand progression which is now bearing Humanity onwards to perfection" (1855, p. 620).

In his letters to the *Nonconformist* and in the first part of *Social Statics*, Spencer cast his eye to God as the promulgator of those providential laws designed to produce the New Jerusalem. Through the remaining parts of his book, however, God gradually receded into the wings, as a more impersonal nature stepped forward. Her actions revealed the same design, even though she relinquished the burning bush and tablets of stone. The nature that took center stage in Spencer's conception would not bend to religious importuning -- no special pleading, no friends at court, not even Laudian High Church incense could move her. Lawful nature replaced a provident God as disposer of progressive advance. The original theological goal of human development remained intact, but would now be insured by fixed natural laws, by the principles of evolutionary progress. Spencer replaced divine providence with natural laws, ultimately to achieve the same moral end.

But that was Spencer, and we think his evolutionary ideas went the way of the dodo. What should we say of Darwin? Well, not, I think what has been suggested by authors like Bowler, Williams, Mayr, and some contributors to this volume, Steve Gould and Will Provine, for instance. I believe that Darwin's evolutionary ideas grew in an environment like to that of Spencer, and that the moral and theological pressures of that environment gave his theory a similar developmental curve. I will consider the history of Darwin's assumptions about progress in three parts, the first dealing with his early notebooks, composed prior to the *Origin of Species*, the second focused on the *Origin* itself and an early draft of it, and the last on the period embracing the composition of the *Descent of Man*. I will argue that from his earliest formulation of the idea of evolution through its mature form in the *Descent of Man*, Darwin conceived of the process as progressive. This progress was marked in several ways. From the early period to the late, Darwin thought that evolution gradually produced ever more complex creatures, that the embryo provided a living paleontological deposit -- an organic picture of this complex development -- and that the most conspicuous instances of evolutionary progress were greater intelligence and a moral sense in the human species.

Ideas of Progress in Darwin's Early Notebooks

In the initial pages of his first transmutation notebook, opened the summer after returning from his five-year Beagle voyage, Darwin speculated that animal groups isolated in a new environment would progressively alter. In July, 1837, he penned the following in his "B Notebook":

> As I have before said, isolate species, especially with some change, probably vary quicker. -- Unknown causes of change. Volcanic island. -- Electricity. Each species changes. Does it progress. Man gains ideas. The simplest cannot help become more complicated; and if we look to first origin, there must be progress. (1960, "B" MS, pp. 17-18)

From the context of his remarks, it is clear that the theory of biological progress over which Darwin's mind played was that of his French predecessor, Jean Baptiste de Lamarck. Shortly after, however, he also entertained Richard Owen's notions about progressive replacement of morphological types, as these jottings reveal:

> Every successive animal is branching upwards different types of
> organisation improving as Owen says simplest coming in and most perfect
> and others occasionally dying out. (1960, "B" MS, p. 19)

In these early pages of his first notebook, Darwin even mused that man was
the goal of progressive evolution:

> Progressive development gives final cause for enormous periods anterior
> to man. Difficult for man to be unprejudiced about self, but considering
> power, extending range, reason and futurity, it does as yet appear . . . .
> (1960, "B" MS, p. 49)

Unfortunately the rest of this passage was cut from the notebook, but the gist
remains: the final cause, the purpose of the long period prior to the human
species's appearance on earth was to allow time for maturation of those
qualities of adaptability, reason, and promise that mark man the most
progressive creature wrought by evolution. Ripeness is all.

Later, in his second transmutation notebook, his "C Notebook," which he
kept from February to July of 1838, Darwin reflected that the animal economy
had conspired to produce human beings, and that if man were suddenly to die
off, new, highly intellectual creatures, much like us, would evolve from current
monkeys -- though the transforming biological pressures would work no more
swiftly nor more obviously than the geological forces Lyell described:

> The believing that monkey would breed (if mankind destroyed) some
> intellectual being though not MAN, -- is as difficult to understand as
> Lyells doctrine of slow movements . . . . What circumstances may have
> been necessary to have made man! Seclusion want &c & perhaps a train
> of animals of hundred generations of species to produce contingents
> proper. -- Present monkeys might not -- but probably would, -- the world
> now being fit, for such an animal -- man, (rude uncivilized man) might
> not have lived when certain other animals were alive, which have
> perished. (Darwin 1960, "C" MS, pp. 74, 79)

Darwin mused in these early jottings that evolving organic relations opened
spaces in the economy of nature that could be filled only by those progressive
qualities distinctive of our species; if human beings failed to occupy this
garden, then another Adam and Eve would, even if they were more simian in

demeanor.

Darwin did, however, allow that some more recently evolved species might have, in a sense, regressed; the individuals of such species might have adapted to new, but more simplified environments, and themselves have become less complex as a result. Mammals migrating back to the sea, for instance, might have lost no-longer-useful appendages. But despite the occasional regression of a species, the higher categories of animals, the genera and orders, would, he thought, continue to display progressive development:

> My idea of propagation almost infers, what we call improvement. All mammalia from one stock, and now that one stock cannot be supposed to be most perfect (according to our ideas of perfection), but intermediate in character. The same reasoning will allow of decrease in character (which perhaps is case with fish, as some of the most perfect kinds the shark. Lived in remotest epochs) . . . . It is another question whether whole scale of Zoology may not be perfecting by change of Mammalia for Reptiles which can only be adaptation to changing world. (1960, "B" MS, pp. 204-05)

Though Darwin often expressed the assumption in his early notebooks that evolution was progressive, yet he handled the idea cautiously. In one often quoted and abused passage he reflected:

> It is absurd to talk of one animal being higher than another -- We consider those where the cerebral structures intellectual faculties most developed, as highest. -- A bee doubtless would where the instincts were. (1960, "B" MS, p. 74)

But this passage merely expresses an insight that is surely true: the standard for "highness" is not a deliverance from nature, but one which we choose. Darwin simply recognized that any standard of measure must be relative to the reasons for selecting it. But he grew ever more certain that man's peculiar endowments -- high intellect and a moral sense -- were quite obviously ideals of perfection, against which other animals could reasonably be evaluated.

Perhaps a more troublesome kind of comment, found frequently enough in Darwin's notes and letters, is of the sort made to Alpheus Hyatt, an American neo-Lamarckian, with whom Darwin had corresponded in the early 1870s:

> After long reflection I cannot avoid the conviction that no innate tendency to progressive development exists, as is now held by so many

able naturalists, and perhaps by yourself. (1902, 2:344)

This was a constant refrain in Darwin's work virtually from the beginning: no innate tendency toward perfection (see also 1909, p. 47). What Darwin objected to, though, was Lamarck's theory that organisms exhibited an *innate* drive toward complexity, toward greater perfection. But this objection to Lamarck does not mean that Darwin rejected the idea that evolution was generally progressive. In his early notebooks, he portrayed progress as the inevitable outcome of the logic of the evolutionary process. Progress was the result, not of an internal drive pushing organisms to perfection, but of an external dynamic pulling them to perfection. He supposed that the environment against which a creature would be selected would be the living environment of other creatures, so that each increase of competitive efficiency, each augmentation of specialization, each new trait evolved to meet a new challenge -- that all these alterations of one individual would call forth reciprocal development in others. The evolutionary situation had a built-in progressive dynamic, a belief in which Darwin frequently expressed, as in this passage from his fourth transmutation notebook, the "E Notebook":

> The enormous number of animals in the world depends on their varied structure & complexity. -- hence as the forms became complicated, they opened *fresh* means of adding to their complexity. -- but yet there is no *necessary* tendency in the simple animals to become complicated although all perhaps will have done so from the new relations caused by the advancing complexity of others. -- It may be said, why should there not be at any time as many species tending to dis-development (some probably always have done so, as the simplest fish), my answer is because, if we begin with the simplest forms & suppose them to have changed, their very changes tend to give rise to others. (1960, "E" MS, p. 95)

Ever alive to the crafty ways of nature, Darwin did recognize, as I have already suggested, that some species seemed to have fallen back, to have been chiseled down to a simpler form. We might then ask, as he did in the continuation to the passage just cited: "Why then has there been a retrograde movement in Cephalopods & fish & reptiles?" "Supposing such be the case," he went on (1960, "E" MS, p. 96), "it proves the law of development in partial classes is far from true." Thus, while the great classes of organisms would generally show improvement, some particular parts might not. But even this partial slippage was, he thought, to be expected -- under the assumption that where it occurred, the underlying simpler animals had been destroyed. In which case, the apparent anomaly would disappear: natural selection inevitably would shave down some complex creatures and shove them into the gap. When the necessary backfilling was completed, however, natural selection would continue to stack species progressively higher in perfection.

The Idea of Progress in the Essay of 1842 and in the Origin of Species

In the *Origin of Species*, Darwin retained his early view of the dynamic and progressive character of evolution. Through constant selection against an ever more complicated and diversified environment, higher creatures would evolve. Darwin thought this process of progressive evolution could explain the more theologically structured belief of most naturalists of his time that earlier creatures were lower in the scale of life, more recent creatures higher. In the *Origin*, Darwin put it this way:

> The inhabitants of each successive period in the world's history have beaten their predecessors in the race for life, and are, in so far, higher in the scale of nature; and this may account for that vague yet ill defined sentiment, felt by many palaeontologists, that organisation on the whole has progressed. (1859, p. 345)

Darwin believed that his theory explained why species more recently laid down in the fossil record were higher than the more ancient. A Biblical fall from grace, wherein animals were supposed to have degenerated as they departed Eden, was thus precluded by the very character of natural selection. But the story of progress was not simply etched in the rocks; like a moving daguerreotype, it could be viewed in the development of the embryo as well.

Darwin observed that the embryos of higher forms resembled those of lower forms, and even conjectured that "the embryonic state of each species and group of species partially show us the structures of their less modified ancient progenitors" (1859, p. 449). He explained the fact and reasonable conjecture, which had previously supported efforts at deciphering the divine plan, by the more mundane operations of natural selection: most progressive adaptations wrought by selection, he argued, would supervene on adults in a variegated environment, rather than on embryos nestled in the stable womb. The ancient, less modified forms of adult animals would, then, resemble more their own embryos than would adults of descendent species resemble theirs. Darwin thought these relationships made it clear "why ancient and extinct forms of life should resemble the embryos of their descendants, -- our existing species" (1859, p. 449). But to further evince this conjecture, he performed several measurement experiments, and reported their results:

> I was told that the foals of cart and race-horses differed as much as the full-grown animals; and this surprised me greatly, as I think it probable that the difference between the two breeds has been wholly caused by selection under domestication; but having had careful measurements made of the dam and of a three-days old colt of a race and heavy cart-horse, I find that the colts have by no means acquired their full amount of proportional differences. (1859, p. 445)

Darwin made like measurements on dogs, pigeons, and the class of animals over which he was master, barnacles. All of these experiments confirmed that the neonates of allied but greatly divergent species had preserved an identity lost in the adults. (Incidentally, the *Origin*, contrary to literate assumption, is rife with Darwin's experiments.)

Darwin's version of what Ernst Haeckel would later promulgate as the biogenetic law suggested that like the fossil record "the embryo comes to be left as a sort of picture, preserved by nature, of the ancient and less modified condition of each animal" (1859, p. 338). Darwin carefully set that small picture against the more majestic Scriptural portrayal of the first days. In the Mosaical beginning, all animals were individually created. But the embryo made "utterly inexplicable" that venerable depiction of "each organic being and each separate organ having been specially created" (1859, p. 480). The embryo could thus be added to the descending strata of fossils as another demonstration of the progressive character of evolution and as another small bomb tucked under a corner of Creationism. (See Gould 1977a, pp. 70-74, for a different interpretation of Darwin's embryology and its relation to the recapitulationist hypothesis.)

Darwin's evolutionary-progressive accounts of the fossil record and embryonic development display his marvelous explanatory strategy in the *Origin*: here, as in many other instances, he was able to explain facts that other naturalists had recognized and, simultaneously, to smite the Creationist alternative. The explanatory power of the *Origin* does not, then, derive from any predictive success of the theory -- of the sort thought necessary by many philosophers for a theory to be accorded scientific status. Rather its power flows from Darwin's craft in weaving into a comprehensible pattern a medley of already recognized facts, facts previously assigned to the incomprehensible ways of the Creator.

These observations about the progressive implications that Darwin drew from his theory do not yet reveal the intimacy of connection between natural selection and progress. Darwin's mechanism of natural selection became animated by the pulse of his profoundly moral conception of nature. The progress of evolution was also moral progress, which infused nature with a higher purpose. This is an essential aspect of Darwin's conception of natural selection unrecognized by his more antagonistic critics as well as by his more recent friends among contemporary historians. To lay bare the moral heart that the mechanical surface of natural selection conceals, we need to consider, if briefly, the history of Darwin's formulation of his principle and its relation to artificial selection.

The Origin of Species insinuates a somewhat misleading relation between artificial selection and natural selection. In the *Origin*, Darwin uses artificial selection as a model for both the articulation of natural selection -- to explain its features -- and also as an argument to convince the reader that these two processes do not essentially differ: if artificial selection exists, so must natural selection. Before historians had carefully scrutinized Darwin's early notebooks, it appeared that the structure of the *Origin* mapped the path by which he had come to discover natural selection -- that is, through analysis of artificial selection. We now think this not to be the case. Only after he had initially formulated the principle of natural selection, did Darwin recognize its analogy with artificial selection. But after he perceived the similarity, I think he did begin to modify his conception of natural selection according to its model, artificial selection. An important feature of that modification was to implant in the mechanism a moral heart.

In an 1842 essay -- a draft that would form the spine of his larger manuscript for the *Origin of Species* -- Darwin first sketched out a coherent outline of his entire theory in which he pictured natural selection as comparable to the work of an infinitely wise being that crafted nature according to a divine plan for what is best. In sentences whose thought jumps about with the energies of literary creation, he reflected:

> If a being infinitely more sagacious than man (not an omniscient creator); during thousands and thousands of years were to select all the variations which tended towards certain ends . . . for instance, if he foresaw a canine animal would be better off, owing to the country producing more hares, if he were longer legged and keener sight -- , greyhound produced . . . . Who, seeing how plants vary in garden, what blind foolish man has done in a few years, will deny an all-seeing being in thousands of years could effect (if the Creator chooses to do so), either by his own direct foresight or by intermediate means, -- which will represent the creator of this universe. (1902, p. 6)

In this essay, natural selection worked as a surrogate for the Creator. The difference, as Darwin conceived it, between man's selection and divinely guided natural selection was that man selected ineptly and for only those qualities that pleased him, while natural selection worked with minute care and for the good of the animal itself. This moral design for the evolutionary process was a direct consequence of Darwin's molding natural selection in the image of the Creator, and of regarding natural selection to be a secondary cause responsive to the primary cause of divine wisdom. Though Darwin's faith in the Biblical Creator waned during the decades after he penned his essay -- so that the *Origin of Species* retains its references to God as much from prudent protection against theologically zealous readers as from his own convictions about a designing Creator -- yet the moral pulse of natural selection remained strong. These moral rhythms become palpable in this passage from the *Origin*, in which Darwin contrasts man's selection with nature's:

> Man can act only on external and visible characters: nature cares nothing for appearances, except in so far as they may be useful to any being. She can act on every internal organ, on every shade of constitutional difference, on the whole machinery of life. Man selects only for his own good; Nature only for that of the being which she tends . . . . It may be said that natural selection is daily and hourly scrutinising, throughout the world, every variation, even the slightest; rejecting that which is bad, preserving and adding up all that is good; silently and insensibly working, whenever and wherever opportunity offers, at the improvement of each organic being in relation to its organic and inorganic conditions of life. (1859, pp. 83, 84)

Darwin's conception of the moral operations of natural selection -- which watched over the fall of a sparrow and numbered the very hairs on the head of man -- supported what was his general sentiment about nature, a sentiment rooted deeply in the Christian tradition, namely that though suffering and death stalked the world, yet all somehow worked to the good. In that most lyrical passage of scientific prose, the final paragraph of the *Origin*, Darwin poignantly expressed his moral view of evolution:

> Thus, from the war of nature, from famine and death, the most exalted object which we are capable of conceiving, namely, the production of the higher animals, directly follows. There is grandeur in this view of life, with its several powers, having been originally breathed into a few forms or into one; and that, whilst this planet has gone cycling on according to the fixed law of gravity, from so simple a beginning endless forms most beautiful and most wonderful have been, and are being, evolved. (1859, p. 490)

Man's Intellectual and Moral Progress

The moral arpeggio of Darwin's theory of evolutionary progress shifted into another key in the *Descent of Man*. The central problem of the first part of the *Descent* was precisely evolutionary progress. Darwin's general theory required progress, particularly in the development of human intellectual and moral faculties, but several critics generally friendly to his theory suggested obstacles preventing such progress (Richards 1987, chaps. 4 and 5).

Alfred Wallace, cofounder of the theory of evolution by natural selection, changed his mind concerning human evolution. He came to deny that natural selection could produce man's big brain and tender moral sentiments. He estimated that, after all, for sheer survival, man required a brain "little superior to that of an ape" (1869, p. 392). For this reason, natural selection, which only operated to produce the sufficient and not the supererogatory, could not explain man's high intellect. Nor, according to Wallace, could it account for our acute moral sensitivity, our altruistic impulses. For moral behavior usually benefits the recipient, not the agent: the fellow who jumps into a river to save a drowning child puts his own life in jeopardy -- something natural selection would not countenance.

These difficulties for the theory were compounded for Darwin by the

observations of William Rathbone Greg, a Scots political thinker who stopped
to reflect on Darwin's theory. Greg pointed out that natural selection seemed
to have a built-in governor, so that in the human line it should disengage
itself. If natural selection produced the social and moral sentiments in man,
Greg argued, such feelings would in protohuman groups prevent the beneficial
culling of the morally and intellectually degenerate. Fueled by the social
instincts that Darwin deemed the foundations of human society, members of a
tribe would prevent their dim-witted friend who wished to pet the sleeping
sabertooth from meeting his natural end. The mentally inferior would live to
procreate another day, so that the numbers of the least advantaged would
increase and thereby produce a regression within evolution. Greg, Scots
gentleman that he was, took the case of the Irish as cautionary:

> The careless, squalid, unaspiring Irishman multiplies like rabbits or
> ephemera: -- the frugal, foreseeing, self-respecting, ambitious Scot, stern
> in his morality, spiritual in his faith, sagacious and disciplined in his
> intelligence, passes his best years in struggle and in celibacy, marries
> late, and leaves few behind . . . . In the eternal 'struggle for existence,'
> it would be the inferior and less favoured race that had prevailed -- and
> prevailed by virtue not of its good qualities but of its faults. (1868, p.
> 361)

In the *Descent of Man*, Darwin thus had two classes of problems bearing
on evolutionary progress that threatened his theory: first, natural selection,
as Wallace maintained, could not produce the high intellect and moral
sentiments that human beings exhibited, since such faculties were in excess of
what was required; and, second, retarding forces produced by natural selection
itself appeared to prevent the further progressive development of the human
species -- our tender mercies would allow the inferior types to outpropagate
the superior.

Darwin ingeniously resolved the first set of problems by applying a device
that he originally developed to explain the traits of the social insects -- the
device of community selection. While altruistic impulse and even high intellect
would little benefit individuals within those tribes of our ancestors -- indeed,
would even be damped down by the governor of individual selection -- yet
such traits would serve the tribe in its competition with other tribes.
Communities themselves would reestablish the struggle for existence on a
higher plane. Within a society, men would no longer struggle individually with

each other for the resources of survival, but between societies, not yet bound
by sentiments of universal brotherhood, the competition would continue just as
effectively, as Darwin explained:

> It must not be forgotten that although a high standard of morality gives
> but a slight or no advantage to each individual man and his children over
> the other men of the same tribe, yet that an advancement in the
> standard of morality and an increase in the number of well-endowed men
> will certainly give an immense advantage to one tribe over another.
> There can be no doubt that a tribe including many members who, from
> possessing in a high degree the spirit of patriotism, fidelity, obedience,
> courage, and sympathy, were always ready to give aid to each other and
> to sacrifice themselves for the common good, would be victorious over
> most other tribes; and this would be natural selection. At all times
> throughout the world tribes have supplanted other tribes; and as morality
> is one element in their success, the standard of morality and the number
> of well-endowed men will thus everywhere tend to rise and increase.
> (1871, 1:166)

If one adds to the device of community selection that more venerable
Lamarckian instrument of the inheritance of the effects of habitual practices
-- an instrument that Darwin never abandoned, even if he did not often deploy
it -- then the first set of problems dissolves.

The other set of difficulties, those suggested by Greg, would not yield so
easily to Darwin's genius. Again, the problem was simply that qualities which
identified the progressive attainments of the human race -- high intellect and
moral sense -- appeared to be swamped out by the vicious, the criminals, and
the imbeciles which inhabited the lower classes of Victorian society. Here
natural selection seemed to be disengaged, since the depressingly worst
appeared to be propagating at a faster rate than the obviously best. In his
analysis, though, Darwin ventured that there were actually inherent checks on
the possible increase among the inferior classes: the poor crowded into towns
would die at a faster rate; the debauched would also suffer higher mortality;
and jailed criminals would not bear children. After careful reflection, it
appeared a just world yet -- fortune would finally snip the thread of the
unworthy. And one could hope, as Darwin and Greg did, that enlightened
social legislation and the impact of moral education would return the
propagatory advantage to the more favorably endowed. Nonetheless, it could
be that civilized nations faced, after reaching a peak, a gradual decline. After

all, as Darwin gloomily observed, "progress is no invariable rule" (1871, 1:177). But he did think, it must be added, that progress was a general rule.

Darwin's worry that his own society might have been slipping into decline makes sense only under the supposition that he expected constant progress, a growth in complexity, finally of a florescence of intellect and moral sense, to be the natural outcome of natural selection. And though he recognized the possibility of devolution, he remained hopeful of continued progress among civilized men. He concluded his consideration of the problem in the *Descent* expressing just that sentiment:

> It is apparently a truer and more cheerful view that progress has been much more general than retrogression; that man has risen, though by slow and interrupted steps, from a lowly condition to the highest standard as yet attained by him in knowledge, morals, and religion. (1871, 1:184)

Conclusion

Among contemporary evolutionary theorists Darwin functions as an icon, an image against which theories may receive approbation or reprobation. To select from the historical Darwin those features that best comport with one's own predilections in the contemporary scientific debate is to have those predilections sanctioned by the master. So, for example, if one dislikes the political and social elitism that seems endorsed by the idea of evolutionary progress, one might claim, as one of our contributors has, "To Darwin, improved meant only 'better designed for an immediate, local environment.'" As Gould goes on:

> Its [natural selection's] Victorian unpopularity, in my view, lay primarily in its denial of general progress as inherent in the workings of evolution. Natural selection is a theory of local adaptation to changing environments. It proposes no perfecting principles, no guarantee of general improvement; in short, no reason for general approbation in a political climate favoring innate progress in nature. (1977b, p. 45)

If we take Darwin whole, we see that his view of progress in evolution does not differ terribly from that of Spencer. Both had their vision of moral and intellectual development sharpened by the Christianity of their youth; and when the vision dimmed, so that the Creator seemed to recede into the vague interstices of nature, yet the moral aspects of the vision were retained. Both believed that evolutionary progress was to be expected, and it would be progress generally in the complexity of organization, and finally in the moral and intellectual faculties characterized by the higher races. The nineteenth century version of evolution may not satisfy our own moral and political demands, but these demands should not obscure either Spencer's or Darwin's theories of evolutionary progress.

Acknowledgment

I am extremely grateful to Richard Burkhardt and Phillip Sloan for aiding in what I suspect they regarded as an attempt to make the worse argument the better.

References

Bain, A. 1863. Letter to Herbert Spencer (November, 17). Athenaeum Collection of Spencer's Correspondence, MS 791, no. 67, University of London Library.

Bowler, P. J. 1986. *Theories of human evolution*. Baltimore: Johns Hopkins University Press.

Carpenter, W. 1841. *Principles of general and comparative physiology*. London: Churchill.

Chambers, R. 1844. *Vestiges of the natural history of creation*. London: Churchill.

Darwin, C. 1859. *On the origin of species*. London: Murray.

Darwin, C. 1871. *Descent of man and selection in relation to sex*. 2 vols. London: Murray.

Darwin, C. 1902. *More letters of Charles Darwin*. Edited by F. Darwin. 2 vols. Cambridge: Cambridge University Press.

Darwin, C. 1909. Essay of 1842. In *Foundations of the origin of species*. Edited by F. Darwin. Cambridge: Cambridge University Press.

Darwin, C. 1960. *Darwin's notebooks on transmutation of species*. Edited by G. de Beer. *Bulletin of the British Museum (Natural History), Historical Series*, 2.

Gould, S. J. 1977a. *Ontogeny and phylogeny*. Cambridge: Harvard University Press.

Gould, S. J. 1977b. *Ever since Darwin*. New York: Norton.

Greg, W. R. 1868. On the failure of "natural selection" in the case of man. *Fraser's Magazine* 78:353-62.

Huxley, T. H. 1902. *Evolution and ethics and other essays*. New York: D. Appleton.

Lyell, C. 1830-1833. *Principles of geology.* 3 vols. London: Murray.

Mayr, E. 1982. *The growth of biological thought.* Cambridge: Cambridge University Press.

Milne-Edwards, H. 1841. *Outlines of anatomy and physiology.* Boston: Little and Brown.

Ospovat, D. 1981. *The development of Darwin's theory.* Cambridge: Cambridge University Press.

Peel, J. D. Y. 1971. *Herbert Spencer: The evolution of a sociologist.* New York: Basic Books.

Richards, R. 1987. *Darwin and the emergence of evolutionary theories of mind and behavior.* Chicago: University of Chicago Press.

Spencer, H. 1842. Letters on the proper sphere of government. *The Nonconformist,* 15 June to 23 November.

Spencer, H. 1851. *Social statics.* London: Chapman.

Spencer, H. 1855. *Principles of psychology.* London: Longman, Brown, Green, and Longmans.

Spencer, H. 1857. Progress: Its law and cause. *Westminster Review,* n.s. 9:445-85.

Spencer, H. 1862. *First principles.* London: Williams and Norgate.

Van Valen, L. 1973. A new evolutionary law. *Evolutionary Theory* 1:1-30.

Wallace, A. R. 1869. Review of *Principles of geology* and *Elements of geology,* by Charles Lyell. *Quarterly Review* 126:359-94.

Williams, G. C. 1966. *Adaptation and natural selection.* Princeton: Princeton University Press.

Orthogenesis and Evolution in the 19th Century: The Idea of Progress in American Neo-Lamarckism

Robert C. Richardson and Thomas C. Kane

"Die Phylogenese ist die mechanische Ursache der Ontogenese."
-- Haeckel, 1874

The theory of natural selection provided the nineteenth century with a thoroughly naturalistic approach to understanding the origin of species. Divergence and specialization, rather than progress, were the hallmarks of *The Origin of Species*. The emergence of alternative theories committed to non-Darwinian mechanisms of evolution might be seen, by contrast, as a return to a theological world view, or as an attempt to accommodate evolution to a theological perspective (Moore 1979; Bowler 1983). Minimally, these non-Darwinian alternatives might accommodate the social vision prevalent in nineteenth century society (Young 1973). In either case, we might expect to find, in these non-Darwinian views, a substantial notion of progress. We show that this expectation is incorrect. While orthogenesis was an integral part of early American Neo-Lamarckism, this involved no substantial notion of progress.

Recapitulation and Progress

In the years following the publication of Darwin's *The Origin of Species*, the principal challenge to Darwinism came in theories which emphasized the orthogenetic character of evolutionary change. According to these views, the critical factor in understanding evolution is the source of variations, and only

149

secondarily the means of their propagation or preservation; indeed, orthogenesis can be defined as *non-adaptive linear evolution* (Bowler 1983, p. 119; Bowler 1984, pp. 233ff.). Few writers maintained a strict orthogenesis for long, but Herbert Spencer, as well as Ernst Haeckel, did embrace orthogenetic views, and correspondingly relegated natural selection to a secondary role in the evolutionary process. Also, both Spencer and Haeckel incorporated a minimal notion of progress. Spencer even *defined* evolution as "the process which is always an integration of matter and dissipation of motion" (1864, p. 286), and says that from "the lowest living forms upwards, the degree of development is marked by the degree in which the several parts constitute a cooperative assemblage" (1864, p. 328).

There is some latitude for incorporating direction to evolution in Darwin, although there is little significance to the notion of progress within his theory. In one of very few comments on progress in the first edition of the *Origin*, Darwin says "all corporeal and mental endowments will tend to progress toward perfection" under the guidance of natural selection (1859, p. 489). In the sixth edition of the *Origin,* Darwin expands on the idea further:

> Natural Selection acts exclusively by the preservation and accumulation of variations, which are beneficial under the organic and inorganic conditions to which each creature is exposed at all periods of life. The ultimate result is that each creature tends to become more and more improved in relation to its conditions. This improvement inevitably leads to the gradual advancement of the organisation of the greater number of living beings throughout the world. (Darwin 1872, p. 116)

Darwin's position was not without ambiguities. At times, he appeared to embrace a recapitulationist doctrine. In *The Descent of Man,* for example, he wrote that there is a "resemblance" between the human embryo and lower adult forms (1871, p. 10). Nonetheless, a nonrecapitulationist principle was the dominant view in Darwin's writings. Darwin finally settled on a standard of "advance in organization" which he derived from von Baer (1828), relying on the degree of differentiation of parts and their functional specialization as a measure of evolutionary advance (see Darwin 1872, p. 117). Darwin wrote in *The Descent of Man* that the human "embryo itself at a very early period can hardly be distinguished from that of other members of the vertebrate kingdom" (1871, p. 9).

This approach permits an interpretation of the evolutionary process as progressive. Nonetheless, Darwin continually insisted that natural selection does not *necessarily* include progressive development, and must be understood as an opportunistic procedure which "only takes advantage of such variations as arise and are beneficial to each creature under its complex relations of life" (Darwin 1872, p. 118). The considered opinion, in our view, was expressed by T. H. Huxley who explained that "so far from a gradual progress towards perfection forming any necessary part of the Darwinian creed, it appears to us that it is perfectly consistent with indefinite persistence in one state, or with a gradual retrogression" (1864, pp. 90-91). Progressive improvement might be accommodated by the Darwinian theory, but is not required by it. The mechanism of natural selection, as well as the secondary mechanisms adopted by Darwin, are consistent with progress, stability, or retrogression. (For a contrary view, defending the position that progress is integral to Darwin's views, see Richards, this volume; for an equally extreme view, but skeptical of *any* commitment on Darwin's part to a useful distinction between "higher" and "lower" forms or to progress, see Grene 1974.)

This is quite unlike a strictly Lamarckian vision, which presents "a necessarily progressive mechanism of evolution" (Bowler 1983, p. 136). According to Lamarck (1809), orthogenesis was the result of an inherent drive toward "perfection" within lineages. In simpler creatures, external sources would produce greater complexity. More advanced organisms tended by their own nature to progress toward a higher level of organization. Reproduction would then preserve these structures. Lamarck's view, like Spencer's, was progressive, with evolution producing more complex structures in place of simpler ones. The inheritance of acquired characteristics, and modification by use and disuse, were little more than corrections geared toward explaining adaptive modifications, while the overall trend of evolution was progressive.

Darwin and his nineteenth century orthogenetic opponents differed fundamentally, not over the progressive character of evolution, but over the *mechanisms* of evolution and the patterns those mechanisms would lead us to anticipate. The non-Darwinian alternatives which were promulgated in the latter half of the nineteenth century promoted an alternative mechanism, resting on endogenous sources of change, and denying the adaptive significance of most evolutionary change. Orthogenesis, so understood, involves some *linear trends in evolution*, but does not necessarily support a *progressive* interpretation of the evolutionary process. Progress is more than pattern. There must be *general* trends -- a sustained overall pattern of change (Ayala, this volume) -- if there is to be progress. For if all trends are local and limited, then the claim that there is progress revealed in the pattern amounts to no more than the claim that there is a pattern. Even the presence of general trends is not sufficient for progress. A progressive trend also requires a "theoretically motivated direction" (Hull, this volume). Neither Darwinism nor its orthogenetic opposition requires a progressive trend in this sense, though both might lead us to expect some general trends. Darwin thought that natural selection would lead to increased specialization (and thereby to speciation). His orthogenetic opponents thought that evolutionary mechanisms would lead to increased complexity. In both cases, it is clear that there is a pattern, even if the patterns are different. In neither case is there necessarily any progress.

Orthogenesis without progress is particularly evident in the case of what Peter Bowler (1983) calls the "American School," or what we term "American Neo-Lamarckism." This included a diverse group of evolutionary theoreticians in the last half of the nineteenth century inspired by Alpheus Hyatt and Edward Drinker Cope. From the end of the Civil War until the turn of the century, the American Neo-Lamarckians provided a vigorous alternative to Darwinism. As the intellectual heirs of Louis Agassiz (1849) and Georges Cuvier (1828), Hyatt (1866) and Cope (1861) began with an orthogenetic view, locating the explanation of evolutionary patterns in the recapitulation of phylogeny by ontogeny, and denying the centrality of adaptationist explanations. By the 1870s, their orthogenesis was compromised in favor of a view emphasizing the effects of use and disuse, together with the inheritance of acquired characteristics. We shall focus our attention on the earliest defenses of American Neo-Lamarckism by Cope and Hyatt in the late 1860s. It is here we find the purest defense of an orthogenetic view. Accordingly, if even here we find no unequivocal commitment to progress, we should expect even less in later expressions of American Neo-Lamarckism. We shall consider these later developments briefly, but we will find this expectation is confirmed. The earlier orthogenetic views have at least a general pattern, if not progress; later views abandon even the pattern, and acknowledge only local trends.

The key to American Neo-Lamarckism is its commitment to the recapitulation of phylogeny by ontogeny. Stephen Gould explained in *Ontogeny and Phylogeny* (1977) that recapitulation rests on two mechanisms: first, *terminal addition* commits us to the view that new generic forms arise by addition to ontogenetic stages in later stages of development (new species arise from the modification of existing forms); second, *acceleration* shortens ontogeny in order to press newly acquired characteristics into earlier periods of development. He writes:

> These principles of terminal addition and acceleration are the preconditions of recapitulation. In progressive evolution, the adult stages of ancestors are crowded back or "accelerated" into the juvenile stages of descendants. Recapitulation is the necessary result of progressive evolution. (Gould 1977, p. 87)

If the origin of genera is to be explained in terms of terminal addition and the subsequent foreshortening of ontogeny by acceleration, then there must be a parallelism between the corresponding ontogenetic stages of allied forms, as well as some level of resemblance between the juvenile forms of more advanced genera and the adult forms of less advanced genera.

Hyatt and Cope thought of this as a "parallelism" of forms varying in completeness, but the product of nonadaptive mechanisms controlling growth and development. The mechanisms they advanced can define progressive trends, in terms of the direction of increasing ontogenetic complexity. We can always stipulate some preferred pattern as progressive. As we shall see, however, this will count as "progress" only in an attenuated sense. Any sense of progress is imposed on a mechanistic theory. Accordingly, it hardly supports a more robust notion of progress than would have been available to the orthodox Darwinian. The mechanisms produce a pattern, but not progress.

Cope on the Origin of Genera

Cope announced the discovery of his law of acceleration and retardation two years before he detailed it in "On the Origin of Genera" in 1868. He maintained that natural selection was important only in the relatively minor modifications characteristic of the differences between species, and that the more profound differences defining genera, and the relations between them, were explicable only in terms of alternative mechanisms. Only differences between species are properly understood as *adaptations*. Differences between genera are a consequence of intrinsic tendencies to vary in predetermined directions independently of adaptive advantages. That is, they are orthogenetic. Cope wrote:

> There are, it appears to us, two laws of [the] means and modes of development. I. The law of acceleration and retardation. II. The law of natural selection .... [While] natural selection operates by the "preservation of the fittest," retardation and acceleration act without any reference to "fitness" at all; ... instead of being controlled by fitness, it is the controller of fitness. Perhaps all the characteristics supposed to mark generalized groups from genera up ... have been evolved under the first mode, combined with some interventions of the second, and ... specific characters or *species* have been evolved by a combination of a lesser degree of the first with a greater degree of the second mode. (Cope 1868, p. 244)

Cope claims that the most closely related genera usually differ from one another only in single characters, but the more distantly related genera differ proportionately to the degree of relationship. Thus, genera are ordered, Cope thought, in the number of steps by which they are removed phylogenetically from one another, and this should generally be reflected in the number of characters by which they differ. In the best case there will be an "exact parallelism" between related genera, so that the differences between the genera will be reflected in the developmental pattern of the higher forms. Thus, if the genera $G_1 \ldots G_n$ differ in characteristics $C_1 \ldots C_n$, then, in *exact parallelism*, the ontogenetic development of the highest form, G_n, should show a sequence of stages exemplifying C_1 through C_n in the same order realized in the phylogenetic series (fig. 1). Cope explains:

> We have then in the embryos of the lower vertebrates at a certain time in the history of each, an "*exact parallelism*" or *identity* with the embryonic condition of the type which progresses to the next degree beyond it, and of all the other types which progress successively to more distant extremes .... The embryo of the fish and that of the reptile and mammal may be said to be generically if not specifically identical up to the point where preparation for the aerial respiration of the latter appears. They each take different lines at this point. (Cope 1868, p. 267)

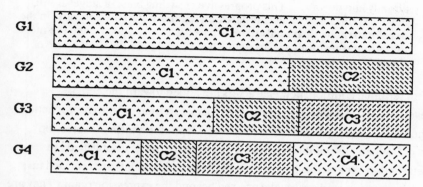

Figure 1. Exact parallelism requires that higher genera replicate the developmental stages of lower genera in the same order, although not necessarily for the same length of time. Thus, the most primitive G1 will differ from G2 in exactly one character, and G2 will exemplify C1 followed by C2. G4, the most advanced form, will exemplify C1 through C4, the distinctive characters of G1 through G4, in that order. Note that in the higher forms each character is accorded increasingly shorter periods.

The less closely allied the genera are, the less complete this parallelism should be. (Cope did not think that there was any considerable parallel at any level, between the most fundamental animal forms, and even the "embryos of the four great branches of the animal kingdom appear to be distinct in essential characters, from their first appearance" (1868, p. 255). Thus, Cope even denies any significant parallel between the embryonic forms of the vertebrates and the invertebrates.)

The cause of this parallelism between phylogeny and ontogeny is found in what Cope calls "acceleration and retardation." Cope, like Haeckel, held that generic characters arise as additions at the later stages of development, and are subsequently accelerated into an earlier developmental period: ontogenetic stages are accorded increasingly shorter periods of time, and later stages are thereby accelerated to an earlier period in the development of the individual. Cope explains that the law of acceleration

> ... consists in a continual crowding backwards of the successive steps of individual development, so that the period of reproduction, while occurring periodically with the change of the year, falls later and later in the life history of the species, conferring upon its offspring features in advance of those possessed by its predecessors, in the line

already laid down . . . . This progressive crowding back of stages is not, however, supposed to have progressed regularly. On the contrary, in the development of all animals there are well-known periods when the most important transitions are accomplished in an incredibly short space of time. (Cope 1868, p. 269)

With terminal addition, new characters are added toward the end of development. Acceleration brings them back to a time prior to reproduction, when they can be transmitted to offspring. Cope calls this the "expression point." When this point is reached, a new generic type is established. In "retardation," characters lying within the "expression point" are retarded in their expression, and eventually pressed beyond the expression point. They are then not inherited. In consequence, the forms of later periods resemble those occurring at earlier ontogenetic periods in their ancestors.

Cope did not think that exact parallelism was an invariant pattern. It was an important pattern, but he knew full well that deviations from it were common. As von Baer had argued in his attacks on recapitulation, Cope knew that a complete morphological correspondence between the adults of "lower" forms and the embryonic stages of "higher" forms was rare. In part, this was due to the modifications in which "one or more characters intervene in the maturity of either the lower or higher genus" (Cope 1868, p. 245). Failures in exact parallelism take two forms. Characters which are common between "higher" and "lower" forms do not always appear in the sequence which their phylogenetic order requires, and sometimes they do not appear at all.

Cope explained this inexact parallelism by appeal to the decoupling of characters in their respective rates of acceleration. Haeckel thought that condensation -- that is, acceleration -- held for the organism as a whole, while Cope thought that acceleration rates could be different for different character sets (cf. Gould 1977, p. 89). By accelerating, or condensing, one character set at a different rate than an independent character set the result is that stages which we would anticipate to be synchronized in development are not, and occur in a developmental order that is anomalous. If C_3 and C_4 belong to distinct character sets and are accelerated at different rates, then their relative order can be inverted. Thus, though some characters would be expected, on phylogenetic grounds, to be synchronous with others in ontogeny, they may be asynchronous. Similarly, although the phylogenetic history might lead us to anticipate a particular order in ontogeny, that order can be

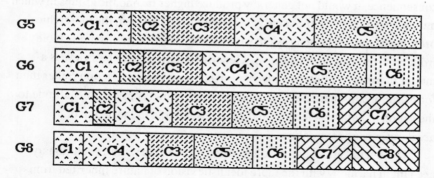

Figure 2. Inexact parallelism allows for less than perfect recapitulation. This can result in the deletion of developmental stages, or in an inversion of order. Although G5 and G6, for example, parallel G1 through G4 of fig. 1 exactly, by G8, the character C2 has been lost. Likewise, by G7 the stages defined by C3 and C4 have undergone an inversion of order.

inverted. Acceleration also invariably leads to the shortening of ontogenetic stages. This means that, with time, stages which were salient in earlier phylogenetic forms may disappear altogether (fig. 2). We should look for parallelism *within* limited character sets, but anticipate no extensive parallelism *between* such sets (Cope 1868, p. 279).

Cope also believed generic characters could change abruptly, and drew an analogy to the striking metamorphosis seen in some insects and frogs. Just as developmental changes may be abrupt, issuing in novel characters, the transformation of genera is often "rapid and abrupt" with long "intervening periods of persistency."

Natural selection, in Cope's view, contributes relatively little to the origin of genera. Cope sees natural selection as "the force which the will of the animal applies to its body, in the search for means of subsistence and protection from injuries, gradually producing those features which are evidently adaptive in their nature" (1868, pp. 289-90). He emphasizes two objections to theories relying primarily on natural selection. First, in the majority of cases, the generic characters serve no adaptive function, and in the remainder, they yield no significant adaptive advantage. In short, generic differences are not adaptive. Second, since natural selection is utilitarian, it would be expected "to have acted differently to produce the same results" (1868, p. 291). As a consequence, it would not generally produce higher taxonomic groups in which multiple species share the same generic character. In short, the hierarchical ordering of species is inconsistent with natural selection, but is predictable from his orthogenetic perspective. Generic similarity, in Cope's view, is a matter of form. A utilitarian mechanism concentrates on function rather than form. In specific characters, by contrast, Cope allows considerable latitude to the action of natural selection, and a diminished role to the action of acceleration and retardation.

The overall effect of this complicated mechanism was an evolutionary account in harmony with the more idealistic vision of nature inherited from Agassiz and, ultimately, the Romantic German *Naturphilosophen*. It emphasized form rather than function. It incorporated a hierarchical ordering of species. Allied genera are ordered in a fashion which, at least roughly, reflects a natural progression of forms. These genera, further, are separated from one another by sharp boundaries. Yet it did this within an evolutionary model.

The distinguishing feature of orthogenetic views is a commitment to an internal mechanism producing an orderly, linear, development over time. Cope's view is surely an orthogenetic view of nature by this standard. Yet the origin of genera hardly displays any progress from lower to higher forms. We have at best an inexact parallelism of forms, further modified by the local changes due to natural selection, and lacking any clear independent criterion for what would ultimately count as progress.

The cause of the pattern, at this stage in Cope's own development, was not explained in terms of mechanical causes, but as a consequence of divine guidance. In Cope's view, there is no naturalistic explanation of the *sources* of the generic characters, and they must consequently be looked for elsewhere. "Our present knowledge," Cope says, "will only permit us to suppose that the resulting and now existing kingdoms and classes of animals and plants were conceived by the Creator according to a plan of his own, according to his pleasure" (Cope 1868, p. 269). As we will see, Cope eventually abandoned the theistic component of this view by accepting that terminal addition was a result of environmentally induced modifications. He did *not*, however, abandon a theological interpretation of evolutionary patterns.

Alpheus Hyatt and Racial Senescence

In 1866, the year that Cope initiated his doctrine of recapitulation, and Haeckel defended it, Alpheus Hyatt also opened a defense of recapitulation and the law of acceleration with a study of fossil cephalopods. Hyatt writes:

> The correlations of these periodical revolutions in the life of the individual with those displayed on a greater scale in the life of the entire order of Tetrabranchiates in time, are wonderfully harmonious and precise. They open a vista through which the individual may be viewed in a new and unexpected light; standing side by side with its own series of forms, it seems to embody the same biological law; not only rising and declining within the narrow limits of its separate existence as they do in their totality, but varying the characteristics of its different periods reciprocally with the more extensive changes of the entire series. (Hyatt 1866, p. 195)

Cope and Hyatt almost certainly inherited the doctrine of recapitulation from Agassiz. Hyatt even says that it is nothing but an evolutionary version of Agassiz' doctrine of embryological recapitulation (Gould 1977, p. 91). For Hyatt, as for Cope, the motive force for evolution is in an intrinsic tendency of species to change. And again like Cope, Hyatt thought the mechanism involved terminal addition and the subsequent acceleration of characters into earlier periods of development. Unlike Cope, Hyatt treated both evolutionary "advance" and evolutionary "decline" as the result of acceleration, together with an internal, preprogrammed, source of modification.

Hyatt was convinced that there is an extensive parallelism between individual development and the modification of species in a lineage. He says the goal of his investigations was

> ... to compare the characteristics of the period of decline of the individual with the adult features of allied species in the hope of finding ... correlations ... between the development of the young and the permanent states of simpler organizations .... [The] life of the individual [cephalopod], so far as the shell and its internal structure indicate what its metamorphoses were, displays during its rise and decline, phenomena correlative with the rise and decline of the collective life of the group to which it immediately belongs. (Hyatt 1866, p. 193)

These parallels he reports are extensive and encompass changes in the numbers

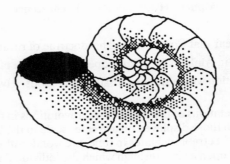

Figure 3. Planospiraled shell with relatively simple septa. The shell grows from the inside out, with the animal occupying only the outer chamber at any given time.

of species as well as variations in form. In the latter category are the ornamentation and the structure of the shells, especially changes in the septa and internal sutures where the septum joins the wall of the shell. In discussing the numbers of species (see Gould et al. 1987), Hyatt distinguishes between the "normal" and "aberrant" nautiloids. The former begin, by his count, with 2 in the Silurian, peak in the Carboniferous with 38 species, and have 2 in the Cretaceous. The latter peak in the Silurian with 114 species and reach extinction by the Cretaceous. Ammonites follow the same pattern but peak in the Jurassic with 200 species, as opposed to 8 for the "normal" and 14 for the "aberrant" nautiloids. The developmental pattern lies in the tendency of individuals, with age, to develop in a planospiral (fig. 3). Hyatt noted that in the period from the Silurian to the Cretaceous the most varied forms of tetrabranchiata come when there are greater numbers of species, and that the most complex forms come later -- though not quite at the end of the lineage. In one subgroup of the nautiloids, Hyatt says the youngest individuals have bent, nonenveloping shells with simple septa, the adult is coiled with an ornamented shell, and with extreme age there is a loss of ornamentation, resulting in some resemblance to the juvenile form. The lineage changes in a parallel way. The most ornamentation and complexity is at the height of the lineage, and the simplest forms occur early and late in its development. (See Erben 1966, for a recent treatment of the evolution of the ammonoids.) The ammonites have an analogous pattern, as Hyatt explained:

... the succession of the four epochs [of expansion in the tetrabranchiata] resembles the succession of the four periods of the individual both in number and their general structural peculiarities.

The first epoch [the Silurian] of the order is especially the era of rounded, and, in the majority of the species, unornamented shells with simple septa; the second [the Carboniferous] is the era of ornamentation, and the septa are steadily complicating; in the third [the Jurassic] the complication of the septa, the ornamentation, and the number of species, about twice that of any other epoch, all combine to make it the zenith of development in the order; the fourth [the Cretaceous] is distinguishable from all the preceding as the era of retrogression in the form and partially in the septa.

The four periods of the individual are similarly arranged and have comparable characteristics. As has been previously stated, the first is smooth and rounded with simple septa; the second tuberculated and the septa more complicated; the third was the only one in which the septa, form, and ornamentation simultaneously attained the climax of individual complication; the fourth, when amounting to anything more important than the loss of a few ornaments, was marked by the retrogression of the whorl to a more tubular aspect, and by the partial degradation of the septa. (Hyatt 1866, p. 200)

For both the order and the individual, there are four stages, each with a characteristic structure: (1) simple septa with rounded shells; (2) shells with more ornamentation and more complex septa; (3) septa, form, and ornamentation at a "zenith" of complexity; and, (4) a degradation of septa and of the general form.

The primary mechanism behind recapitulation, for Hyatt and Cope, was the law of acceleration. Cope thought degeneration of form was the result of retardation and the subsequent loss of characters. Hyatt construed both advance and decline -- both progressive and regressive evolution -- as a result of acceleration. Senile features were the analogues of juvenile features, occasionally detectable in pathological specimens at the zenith of a lineage, but never present in the juvenile stages (Hyatt 1866, p. 201). A lineage begins as a juvenile form, and, with maturity, adds characters. These added characters are accelerated to earlier periods, while more characters are added. Finally, senile features are added and subsequently accelerated, as are all terminal additions. Eventually, all intermediate stages are crowded out, and the result is virtually indistinguishable from a juvenile form. Hyatt explains:

... there is an unceasing concentration of the adult characteristics

of lower species in the young of higher species, and a consequent
displacement of other embryonic features which had themselves, also,
previously belonged to the adult periods of still lower forms.

This law applied to such groups as have been mentioned, produces a
steady upward advance of complication. The adult differences of the
individuals or species being absorbed into the young of succeeding
species, these last must necessarily add to them by growth greater
differences, which in turn become embryonic, and so on; but when the
same law acts upon some series whose individuals alter the shell in old
age, precisely the reverse occurs, and a general decline takes place. The
old age characteristics, in due course of time or structure, become
embryonic, and finally affect the entire growth and aspect of the higher
members of the series. (Hyatt 1866, p. 203)

Assuming the universality of the law of acceleration, characters which develop
later in ontogeny will inevitably shift to earlier periods of development. In
earlier phylogenetic stages, this will lead to an increase in complexity, to "a
steady upward advance of the complication" (loc. cit.). In later stages, the
senile characters also accelerate to earlier periods of development, and the
senile characteristics of the lineage become embryonic stages in the individual.
This "racial senescence" depends finally on what characters appear in the
ontogeny of more advanced forms. Hyatt thought that the "vital powers"
which allowed a lineage to flourish must eventually be exhausted, leaving it
unable to maintain or improve its position in the natural order. With the
exhaustion of these "vital powers," new senile characters appeared. In the
order, as in the individual, there is a predetermined tendency toward
exhaustion, degeneration, decay, and extinction.

Hyatt, like Cope, maintained that the development of a lineage paralleled
the development of the individual. This meant that, at later phylogenetic
periods, there would be decay and, finally, death. Cope thought the periods of
decay were the result of attenuation of the developmental stages and
subsequent loss of characters. Hyatt, on the other hand, thought the periods
of decay were the result of acceleration of senile stages, with a corresponding
loss of intermediate developmental stages. In Hyatt, no less than Cope, we
have both an orthogenetic vision and a clear general pattern. In neither case
do we have a tractable case of evolutionary progress.

Regressive Evolution

Cave organisms provided a useful paradigm for American Neo-Lamarckism (Kane and Richardson 1985). Alpheus Packard (1839-1905), another product of the Agassiz school and a collaborator of Hyatt and Cope, remarked that "the main interest in the ... studies on cave life center in the obvious bearing on the facts upon the theory of descent" (1888, p. 116). The morphological pattern of troglobites is striking and simple. Four characteristics are especially noteworthy: (1) a general loss of pigment; (2) a diminished size of eyes, with a loss of visual ability; (3) an attenuation of appendages, with hypertrophy of other nonoptic sensory organs; and (4) a reduced metabolic rate. The troglomorphic pattern is common but not universal. Certain species of the scavenger beetle *Ptomaphagus* have eyes and pigments. The Mexican cave fish *Astyanax mexicanus* exhibits considerable diversity of form. Other examples fit the pattern even less well (Culver 1982). Some of these troglomorphic characteristics are likely to be adaptations, such as the reduced metabolic rate. Others, such as the loss of pigment, are less likely to be so. The principal problem facing American Neo-Lamarckians was to explain the apparently nonadaptive features of cave organisms.

Cope (1872) explained eye loss in the cave fish *Typhlichthys* and the cave crayfish *Orconectes* (fig. 4) in terms of retardation of growth, although the retardation was environmentally induced. Cave animals, in successive generations, did not develop to the same level of complexity as their ancestors. There is a loss of characters, especially those arising late in development and which are correspondingly relatively recent phylogenetic novelties. Cope thought that the attenuated growth rate resulting from retardation would at least diminish the size of the eyes and eliminate other characters; but, since these are structures that arise relatively early in development, retardation would not, by itself, account for the *disappearance* of the visual organs. Retardation could reduce, but not eliminate, eyes. Without some additional impetus, they would persist into maturity. Cope (1872, p. 417) concluded that the loss of such structures must be a consequence of "retardation and atrophy." The attenuation of growth in retardation could initiate a reduction in the size of the eye that could be exaggerated in subsequent generations. This attenuation would lead to dysfunction and atrophy of organs, and disuse

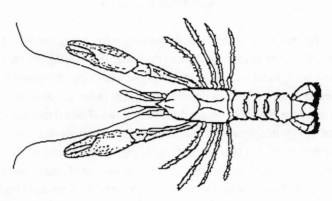

Figure 4. *Orconectes inermis* Cope. Adapted from Cope 1872.

would result in their elimination.

Once admitted, the appeal to environmental effects was bound to persist. In one of the earliest systematic studies of cave fauna, Packard reasons as follows:

> ... it seems evident that geologically speaking the species were *suddenly* formed .... [The] comparatively sudden creation of these cave animals affords, it seems to us, a very strong argument for the theory of Cope and Hyatt of creation by acceleration and retardation .... The strongly marked characters which separate these animals from their allies in the sunlight, are just those fitting them for their cave life and those which we would imagine would be the first to be acquired by them on being removed from their normal habitat. (Packard 1871, p. 759)

Packard suggests that if a locust were introduced into a cave habitat, in which it must live "by clinging to the walls," its legs would grow longer and its antennae would elongate (fig. 5). Coloration and eyes would prove useless and be reduced. The favored explanations for troglomorphic traits in the rising Neo-Lamarckian School were the environmentally induced atrophy with subsequent loss of the organs, or degeneration from "disuse," and an environmentally induced increase in organs, or accentuation with "use." The motive force for evolutionary change became exogenous rather than endogenous.

It is easy to reconcile the appeal to use and disuse with the "law of

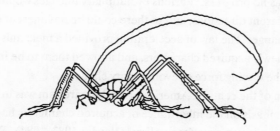

Figure 5. *Raphidophora subterranea*. Adapted from Packard 1871.

acceleration and retardation" originally defended by Cope and Hyatt. Most significant for our purposes is that this effectively compromises the orthogenetic emphasis of the earlier views. The critical influence is now environmental, and the modifications are seen as responses, even if not as adaptations, to the cave environment. Where local modification is the focus, any semblance of a notion of progress vanishes. Pattern is lost. Variation rather than direction becomes the rule.

The Decline of American Neo-Lamarckism

As work in the American School progressed, it increasingly emphasized the adaptive character of evolutionary change. Appeals to use and disuse, and to the inheritance of acquired characters, became the leitmotif of American Neo-Lamarckism. In his paleontological work, Cope concentrated on the effects of use as an adaptive agent of evolutionary change, and used it to understand the evolutionary trends linking various fossil specimens. He saw these trends as incompatible with Darwinism. Selection on random variations should produce an irregular pattern, rather than the steady trends he found in the record. A favorite example of Cope's was the development of the horse from its five-toed ancestor, showing the size increase in the descendants within a lineage.

Hyatt admitted that evolutionary developments often involved utilitarian modifications, and he too thought that the patterns in the fossil record were evidence against a Darwinian view. Yet, although both Cope and Hyatt saw trends, there was no progress. Various evolutionary lineages would be modified in different directions, and so there could be no inherent direction to evolutionary change. The law of acceleration provided a mechanism for incorporating newly acquired characters, and allowed them to be inherited. It was no longer the driving force of evolutionary change.

By the turn of the century, American Neo-Lamarckism was in retreat. Perhaps the emphasis on the inheritance of acquired characters had been compromised by a "rising experimentalism" (Bowler 1983, 1984). Perhaps the rise of Mendelism took its toll. Neither of these can be the full story, since the decline of American Neo-Lamarckism began before either could exact its price; moreover, the inheritance of acquired characters survived both. Whatever the cause of its decline, the response of Cope's disciples, William Berryman Scott (1917) and Henry Fairfield Osborn (1907), was to turn away from the inheritance of acquired characteristics and explain the trends in the fossil record with a purely orthogenetic mechanism. The movement had come full circle. Having begun with an emphasis on orthogenesis, and subsequently accommodating exogenous sources of evolutionary change, American Neo-Lamarckism, in its own period of senile decay, ended with orthogenesis as well.

References

Agassiz, L. 1849. *Twelve lectures on comparative embryology.* Boston: Henry Flanders.

Baer, K. E. von. 1828. *Entwicklungsgeschichte der Thiere: Beobachtung und Reflexion.* Königsberg: Bornträger.

Bowler, P. J. 1983. *The eclipse of Darwinism.* Baltimore: The Johns Hopkins University Press.

Bowler, P. J. 1984. *Evolution: The history of an idea.* Berkeley: University of California Press.

Cope, E. D. 1866. On the Cyprinidae of Pennsylvania. *Trans. Am. Phil. Soc.* 13:351-99.

Cope, E. D. 1868. On the origin of genera. *Proc. Acad. Nat. Sci. Phila.* 20:242-300.

Cope, E. D. 1872. On the Wyandotte Cave and its fauna. *American Naturalist* 6:406-22.

Culver, D. C. 1982. *Cave life.* Cambridge: Harvard University Press.

Cuvier, G. 1828. *Le règne animal distribúe d'après son organisation.* Paris: Fortin.

Darwin, C. 1859. *On the origin of species.* First Edition. London: John Murray.

Darwin, C. 1871. *The descent of man*. Second Edition. New York: Clark, Given, and Hooper.

Darwin, C. 1872. *The origin of species*. 6th Edition. London: John Murray.

Erben, H. K. 1966. On the origin of the ammonoids. *Biol. Rev.* 41:641-58.

Gould, S. J. 1977. *Ontogeny and phylogeny*. Cambridge: Harvard University Press.

Gould, S. J., N. L. Gilinsky, and R. Z. German. 1987. Asymmetry of lineages and the direction of evolutionary time. *Science* 236:1437-41.

Grene, M. 1974. *The understanding of nature*. Dordrecht: D. Reidel.

Haeckel, E. 1874. *Anthropogenie: Keimes- und Stammes-Geschichte des Menschen*. Leipzig: W. Engelmann.

Huxley, T. H. 1864. Criticisms on "The origin of species." In *Collected essays*, T. H. Huxley. Vol. II. New York: Macmillan.

Hyatt, A. 1866. On the parallelism between the different stages of life in the individual and those in the entire group of the Molluscous order Tetrabranchiata. *Mem. Boston Society of Nat. Hist.* I, Part 1:193-209.

Kane, T. C., and Richardson, R. 1985. Regressive evolution: An historical perspective. *The NSS Bulletin: Journal of Caves and Karst Studies* 47:71-77.

Lamarck, J.-B. 1809. *Philosophie zoologique*. Paris: Dentu.

Moore, J. R. 1979. *The post-Darwinian controversies*. Cambridge: Cambridge University Press.

Osborn, H. F. 1907. *The evolution of mammalian molar teeth to and from the triangular type*. New York: Macmillan.

Packard, A. S. 1871. The Mammoth Cave and its fauna: On the crustaceans and insects. *American Naturalist* 5:744-61.

Packard, A. S. 1888. The cave fauna of North America, with remarks on the anatomy of the brain and the origin of the blind species. *Mem. Nat. Acad. of Science (USA)* 4:1-156.

Scott, W. B. 1917. *The theory of evolution: With special reference to the evidence upon which it is founded*. New York: Macmillan.

Spencer, H. 1864. *First principles*. New York: Appleton & Co.

Young, R. M. 1973. The historiographic and ideological contexts of the nineteenth-century debate on man's place in nature. In *Changing perspectives in the history of science*, ed. M. Teich and R. Young. London: Heinemann.

The Concept of Progress in Cultural Evolution

Robert C. Dunnell

It is certainly appropriate to consider archaeology and anthropology in a book assaying the role of "progress" in evolutionary thought. Progress, to an extent not true in biological evolution, has been a central focus of evolutionary thought in anthropology. From the outset it is crucial to realize that "evolution" as it has been used in the social sciences for more than a century is a fundamentally different kind of explanatory system than "evolution" in the natural sciences. Apart from the casual use of "evolution" as a synonym for change, it almost invariably carries a connotation of progress in one sense or another. It is my intent here to elucidate the role "progress" plays as a concept in anthropology, especially in archaeology. Because of the contrastive anthropological and natural science meanings of evolution, it is important to distinguish Cultural Evolution as a set of social science explanatory systems from cultural evolution in the natural science sense of the evolution of cultural phenomena.

Cultural Evolution is not, as is sometimes still supposed (Yoffee 1979), the application of biological evolution to cultural phenomena. This latter possibility is just now beginning to be realized in anthropology (Boyd and Richerson 1985; Cavalli-Sforza and Feldman 1981; Rindos 1980, 1984). These initial efforts are largely programmatic and attempt to demonstrate the applicability of a Darwinian model to sociocultural phenomena (primarily social phenomena) generally. To a greater (Boyd and Richerson 1985) or lesser (Rindos 1980) extent, these efforts embody conceptual compromises with their origins in Cultural Evolution and for that reason an acquaintance with the Cultural Evolution methodology is useful in understanding their structure and conceptual content. None of these excursions into the application of biological evolution to cultural phenomena has, however, addressed the role of "progress"

169

in any detail. Consequently, I will not consider this particular brand of cultural evolution further in spite of my sympathy with the endeavor.

Because of the historical confusion surrounding the meaning of evolution in anthropology and in the natural sciences, I shall briefly sketch the differences between them. Since these differences have been discussed elsewhere in detail (Blute 1979; Dunnell 1980, in press), an abbreviated account suffices to set the stage for an examination of the contrastive statuses of the notion of progress in each. Whereas progress in biological evolution is a description, perhaps an empirical generalization, about history, I hope to show that in Cultural Evolution progress is not only an observation about history but it is also a *cause* of that history. The liabilities of Cultural Evolution as science have been recognized sporadically within anthropology for a very long time (Boas 1896, 1911; Steward 1955; Willey 1961). Today, Cultural Evolution is a minor element of sociocultural anthropology (Stocking 1987, p. 328); however, it continues, both in frank expression and in various guises, to guide a considerable amount of archaeological thinking. In part, this anomaly has a methodological origin. When archaeologists embraced explicitly anthropological goals in the 1960s, they were compelled to draw conclusions about things not in evidence in the archaeological record (e.g., kinship, social structure). While alternative methodologies were developed (Binford 1968; Watson et al. 1971), Cultural Evolution remained a convenient, if not unchallenged (e.g., Binford 1968, 1983; Dunnell 1982; Freeman 1968; MacWhite 1956; Slotkin 1952) device for projecting behavioral phenomena into the past, paralleling the conjectural histories of the nineteenth century. At a deeper level, however, one supposes that Cultural Evolution has continued in use partly because Darwinian evolution is seen as incapable of explaining the major "facts" of cultural history as presently known. Consequently, I conclude by examining two of the troublesome "facts" from a selectionist perspective. While this examination yields no conclusive results, it does, I hope, show that a prior rejection of a Darwinian approach is not justified and that the appearance of progress is potentially explicable in such a framework.

Biological and Cultural Evolution

Even to the casual reader, Cultural Evolution and biological evolution are different, if only because of the language of discourse and the typical subjects. As pervasive as terminological differences are, they do not necessarily signal a fundamental contrast in methodology. The simple effects of different scholarly traditions might be at work as seemingly implied by those who regard Cultural Evolution as an application of biological theory to cultural phenomena. Yet, it is easy enough to demonstrate that cultural evolution and biological evolution do not share the same metaphysical view of the nature of reality and, as a consequence, have developed radically different methodologies. Apart from the term evolution itself, they share only the claim that they explain the contemporary world by reference to its history.

Cultural Evolution was the dominant methodology of nineteenth century anthropology. In the early twentieth century it receded from the forefront of the discipline only to reassert itself again, beginning in the 1940s. Throughout this long period, however, no particular version of Cultural Evolution gained the kind of ascendancy as to be characterized as a paradigm (Kuhn 1962; cf. Stocking 1987, pp. xiii-xiv). Nearly every writer presents a view of Cultural Evolution that is to some degree unique. In a brief undertaking it is therefore necessary to focus on the recurrent tenets that link variable research programs together and set them apart from other anthropological approaches. Unavoidably, some significant texture is lost in the process. Because a period of 40 years separates the nineteenth and twentieth century periods of active use and development, some of the variability of Cultural Evolution is retained by describing each separately. While there is continuity in the role of progress across this hiatus, the importance of progress to Cultural Evolution and its specific interpretation is different in the two periods.

Nineteenth Century Cultural Evolution

Cultural Evolution as a methodology is usually traced to the works of Spencer (esp. 1880) in the latter half of the nineteenth century. The stature of Spencer as a social philosopher and the modern character of his writings overshadow the fact that he was but one participant in a major intellectual reorientation that took place in the 1860s. Further, the reorientation was built with notions, including that of progress, that had much longer histories in Western thought, particularly in eighteenth century progressivism (Stocking 1987). An anthropological Cultural Evolution was first forwarded by others in Spencer's intellectual circle, most notably Lane Fox (later Pitt Rivers) (1875a, 1875b), Lubbock (later Lord Avebury) (1865, 1870), McLennan (1865) and Tylor (1865, 1871, 1881). Linked by a rejection, sometimes implicit and sometimes explicit, of theological orthodoxy, they took humankind to be a psychic unity governed by laws that were extendable into the past by means of a Lyellian uniformitarianism. Those laws were objectified in the development of humankind from lower to higher form, technologically, in control over nature, and most especially, morally. Modern "savages" were seen as arrested examples (the principle of survivals) of the universal process of development. While archaeological evidences were considered to greater (Lubbock) and lesser (McLennan) degrees, reflecting the gradual emergence of archaeology as a distinct kind of inquiry, the focus of attention for and motivations of Cultural Evolutionists lay squarely in contemporary human society. Finally, the interest of the Cultural Evolutionists lay in explaining the process (culture) rather than in explaining the diversity of forms (cultures).

Stocking (1987) sees the crystallization of Cultural Evolution as arising from the impact of Darwin on Victorian intellectuals. There can be no doubt that the general intellectual climate created by the acceptance of Darwinism and its implications for people (Huxley 1863) was conducive to the development of Cultural Evolution, particularly in providing a nontheological rationale for the unity of humankind and in dismissing the degenerational view of change that had truncated late eighteenth century progressivism. This general connection, however, did not lead to any methodological similarity to Darwinism as witnessed by the key tenets of Cultural Evolution just reviewed. It was the establishment in 1859 (Grayson 1983) of a greatly expanded human antiquity with the acceptance of the work of Boucher de Perthes on geological grounds (Lyell 1863) that made the whole approach plausible to midnineteenth century intellectuals. Cultural Evolution served to rationalize the differences among classes in British society, the differences among peoples that Western exploration and colonization were bringing to light, and the rapidly developing archaeological record of Europe. Although Lubbock must certainly be counted among the founders of archaeology, the impact of Cultural Evolution in archaeology is considerably more tenuous than is often depicted today, and archaeological materials played a role different from that played by observations about contemporary people.

Archaeology did not begin to emerge from natural history as a whole until the 1840s, specifically with the so-called "Three Age System" of the Danish archaeologist Thomsen -- the familiar Stone, Bronze, and Iron ages (Daniel 1962). The recognition of artifacts as works of human beings and the appreciation that different kinds of artifacts might reflect chronology were current notions from the eighteenth century, but Thomsen formalized these notions as a device for the organization of museum specimens and his scheme found immediate favor elsewhere, including Britain (Wilson 1851). Thomsen, as had his predecessors, *assumed* that the Iron Age assemblages were the most recent because they most closely resembled the contemporary technology of Western Europe; chronology was reflected in increasing technical sophistication. Plainly, the notion of progress, drawn from the same cultural milieu that spawned Cultural Evolution, was at work as a chronological tool. Thomsen's Danish colleague, Worsaae, on the model of geological thought, conducted stratigraphic excavations that confirmed empirically Thomsen's assumed chronology and the apparent integrity of his stages long before the geological antiquity of people was firmly established. But as the human chronology expanded, the fact that glacial age archaeological assemblages were exclusively members of the Stone Age further confirmed the main outlines of the Thomsen scheme. Additional excavation lead to refinements of the system, first by the division of the Stone Age into the familiar Paleolithic and Neolithic (Lubbock 1865). By the end of the nineteenth century, many "cultures" had been named and recognized on the basis of stratigraphic excavation, principally in French rockshelters. While stages still served as a framework, quite subtly the ordering principle that had begun as a generalized notion of technological progress had been replaced with a reliance on stratigraphic association and order. Archaeology has embarked on a different methodological course from that of anthropology.

Anthropology continued to be dominated by Cultural Evolution throughout the nineteenth century. Contemporary peoples were assigned to stages

(Morgan's [1877] Savagery, Barbarism, Civilized), comparable to the stages of Thomsen, and through which it was supposed the ancestors of Europeans had passed. While anthropologists continued to draw upon archaeological results, the archaeological record was not so much explained by Cultural Evolution as it served to provide an empirical warrant for believing in the validity of Cultural Evolution and its application to the contemporary world. The lack of detailed interaction with archaeology no doubt stemmed in part from the interest of Cultural Evolutionists in kinship, marriage forms, and law, elements that could not be studied directly in the archaeological record. The survivals represented by modern primitive peoples presented a certain problem. Why did they persist in unaltered form? Climate and geography had figured heavily in accounting for survivals earlier in the century, but in the late nineteenth century inherent racial differences came to dominate. Thus, while stages may have begun as analytical categories imposed on a continuum of development, by the late nineteenth century they had acquired empirical status. Advanced traits when found in otherwise savage contexts were explained as the result of contact with Europeans rather than being taken as evidence of developmental continuity (Stocking 1987).

To appreciate the implications of this use of stages, one has to inquire into what the nineteenth century Cultural Evolutionists took to be the cause of this developmental sequence of stages. The ultimate cause, of course, was a natural law of progress, a simple rephrasing of the observations themselves. One looks in vain for selection as a mechanism in a nontrivial role. Rather, human intention, knowledge, intelligence, and desire were accorded the role of causal agent, almost without second thought. This rationale, of course, was borrowed from contemporary common sense that placed human fates in human hands. Selection was conceptualized as "survival of the fittest," but fittest was defined by the standard of contemporary European civilization; selection was simply the removal of the less fit, be they customs or peoples, by the more fit.

North Americans shared many of the same cultural values with their Western European counterparts, and Morgan, a major figure in Cultural Evolution was an American, but Cultural Evolution never gained the same currency in academic circles here. In North America, as in Australia (Mulvaney 1977), the archaeological record did not indicate a long period of development (or retardation for that matter) comparable to that of Europe and it directly challenged the universality of the developmental process and archaeological stages central to Cultural Evolution. The increasingly strong racist implications, the empirical difficulties presented by areas outside western Europe, along with methodological critiques (Boas 1896, 1911) led to the general abandonment of Cultural Evolution in the early years of the twentieth century. Indeed, the early twentieth century was the heyday of ahistorical science and all historical sciences suffered, at least in comparison.

There are a few observations about nineteenth century Cultural Evolution that are pertinent. First, it plainly embraced an essentialist (Hull 1965; Sober 1980) or typological (Mayr 1959) metaphysical view of reality. The past, on the model of the present, was conceptualized in discrete packages (stages, cultures) in which only differences between them were significant. Variation within these units was without developmental significance. This is nowhere plainer than in the reaction to the growth of knowledge about the archaeological record itself. New data were either assigned as members of an extant stage or a new unit was created to accommodate them. In short, the relation between units in nineteenth century Cultural Evolution was one of difference; change had no conceptual role in the structure. The past was treated as just another place, one separated from modern western Europe by years rather than by miles. It is little wonder that from the outset Darwin attempted to distance his theory from that of Spencerian evolution (Gould 1977). As many analysts have emphasized (Gould 1977; Hull 1965; Lewontin 1974; Mayr 1982), Darwin's principal contribution was the materialist ontology that permitted the conceptualization of directional change, bringing the study of change within the purview of science.

Because of this essentialist metaphysic, Cultural Evolution could not accord selection a significant role in shaping developmental relations. The variation required by a selective mechanism has been eliminated a priori as noise. The stages of Cultural Evolution were not analytical devices; they were accorded empirical status. The relations between succeeding stages were ones of transformation (Sober 1980). Development was a matter of elevating one's self to the next stage by acquiring the characteristics of the higher stage. This transformation was accomplished not by the operation of any mechanism but by the desires of the actors themselves. The methodology of Cultural

Evolution had been thoroughly conflated with its data.

However significant their effects may be, these differences are subtle in practice, and nineteenth century Cultural Evolutionists firmly believed that they were operating within the kind of theory espoused by Darwin.

Twentieth Century Cultural Evolution

Cultural Evolution was revived in the midtwentieth century. In North America, this development was stimulated by the work of Leslie White (1949, 1959) and was a phenomenon of sociocultural anthropology. Archaeologists rejected this formulation initially (Willey 1961) even though they were using a generic Cultural Evolution approach complete with stages (Willey and Phillips 1958). Their stages (e.g., Formative, Classic, Post Classic), however, lacked the progressive terminology of the new Cultural Evolution and were regarded as empirical generalizations about the course of history rather than explanatory in and of themselves (Willey 1966, p. 4). It is ironic, though thoroughly understandable given the different trajectories followed by European and American anthropology, that Cultural Evolution was revived about the same time in Europe but in archaeology (Childe 1935, 1946) and not in sociocultural anthropology.

Although sharing many features, including an essentialistic metaphysic, with its nineteenth century predecessor, the new Cultural Evolution departed from it in a number of ways. First, White and others (e.g., Carneiro 1967) sought explicit precedent in Spencer and realized that Cultural Evolution and their revitalization of it were different from Darwinian evolution. They did not analyze the nature of the differences, but this did not preclude claims that biologists, in following the Darwinian model, had missed all of the important stuff of history, and were, in effect, wrong (Carneiro 1972; Sahlins 1960). Inasmuch as the architects of the twentieth century Cultural Evolution were explicit about the differences, the conflation of biological and Cultural Evolution that characterized the late 1960s and 1970s in American anthropology and archaeology arose from poor scholarship.

Second, the nineteenth century Cultural Evolutionists had construed progress in rather unspecific terms, as betterment or similarity to modern

European culture. Certainly, "crudeness" and "primitiveness" in a technological sense were occasionally used to infer order where other chronological indicators were unavailable; however, the core was a succession of stratigraphically defined assemblage types. For White and his successors, progress took on a more specific interpretation: progress is the increase in the amount of energy captured by society. The most primitive societies used the least energy whereas contemporary Western cultures used the most. Allowing that direct measurement of energy use is a difficult business, sociocultural and political complexity were generally adopted as surrogates for energy usage. The stages employed by the new Cultural Evolutionists, the familiar Band, Tribe, Chiefdom, State sequence (Sahlins 1972; Service 1962, 1975; cf. Fried 1967) had the same structure as the nineteenth century analog but it was no longer closely tied to technological or other attributes readily assessed in archaeological data.

Indeed, and in contrast to nineteenth century expressions of Cultural Evolution, the new Cultural Evolution typically lacked any connection to the existing archaeological evidence of cultural development. Although the new Cultural Evolutionists *asserted* that human cultural development had followed this course from simple to complex, its foundations lay squarely in the categorization of ethnographic societies. Recently, Leonard and Jones (1987) have shown that the stages that are the "quanta" of Cultural Evolution do not have an empirical warrant even among ethnographic groups. Using traits commonly employed in assigning groups to a particular stage, they examined the occurrence of these traits using *Ethnographic Atlas* (Murdoch 1967) societies. Rather than finding definite clusters of these traits which might provide an empirical basis for Cultural Evolution stages, they found these traits to be distributed clinically, without suggestion of inflections in the number of societies with particular suites of traits that might be taken to represent stages.

The work of Robert Carneiro (1968, 1970) provides insight into this aspect of the new Cultural Evolution. I pick Carneiro's work, not because it is flawed in some way that suits my purposes here, but on the contrary, because it represents some of the more thoughtful work within the new Cultural Evolution. He is one of a very few Cultural Evolutionists to attempt to give the model specific, nonanecdotal, empirical interpretation.

In 1968 Carneiro proposed that "the degree of regularity in the relative order of development of any two traits is directly proportional to the evolutionary distance between them" (Carneiro 1968, p. 363), a proposition that he identifies as an evolutionary "law" (Carneiro 1970, p. 494) in a debate with Bayard (1969). It is easy to show that the "law" is an empirical generalization about a specific set of data. For example, the key notion of *evolutionary distance* is defined by Carneiro as the number of societies in which the higher trait of a pair appears without the lower trait. Inasmuch as this number would change with the sample of societies examined, it is clearly tied to a specific set of observations rather than to a theoretical principle.

The generalization is based on a scaling of 100 societies using 50 traits (Carneiro 1968, p. 354), itself a revision of a larger matrix (Carneiro and Tobias 1963). Both "society" and "trait" are essentialist entities, a point which is unexamined by Carneiro. Setting aside the empirical reality of societies and looking only at the trait units, there are a number of problems. Bayard (1969) is critical of their definition (e.g., how do you recognize "State regulation of commerce" or "corvée"), but as Carneiro (1970) points out, he has gone to some lengths to make the definitions clear. More to the point than the identification of a given trait is the reason why Carneiro chose these particular traits to describe societies as opposed to any other particular formulations equally well defined. Carneiro's traits are, by and large, traditional *ad hoc* descriptive categories. Now if any *ad hoc* set of descriptive terms displayed the property set out in Carneiro's law, there might be reason for pause, but this is not the case. Only "scalable" traits (Carneiro 1968, p. 368) may be used, i.e., traits with asymmetrical combinatorial properties such that in any pairwise comparison the "higher" trait must always occur in the presence of the "lower," but the lower may appear both in the presence of the higher and by itself. Thus, it is apparent that increasing complexity or progress is not discovered by scaling societies, but rather it is built in a priori by trait selection. Carneiro's law is a mechanical corollary of his initial assumption of progress. Indeed, the law is a mechanical property of all cumulative lists.

What is the basis of the claim that this generalization is evolutionary? Carneiro argues that "if scale analysis reveals a distinct regularity in the *serial* arrangement of *certain* traits" (1968, p. 335, emphasis added), then chance cannot be responsible and an "orderly process" is indicated. Carneiro then simply asserts " this 'orderly process' is nothing less than evolution" (1968, p. 355). As already shown, the selection of "certain" traits produces the pattern observed; the claim that the pattern is evolutionary would appear to hinge on the "serial arrangement" clause, the implication clearly being that serial arrangement reflects chronology of trait development. But this is manifestly not the case. Carneiro's societies are, with two exceptions, more or less contemporary. The serial arrangement derives from scaling certain traits and not from independent data on the times at which particular traits appear. In fact, had Carneiro been faithful to his data in this regard, he would have been forced to conclude that advanced traits appear before primitive ones because both of his premodern cases are complex civilizations, not primitive societies. Carneiro recognizes this problem explicitly, and so "tests" his proposition by looking at the history of Anglo-Saxon society where he finds the order largely confirmed, a result not wholly unexpected inasmuch as the traits used to describe the Anglo-Saxon sequence were selected because of their scalability. Carneiro is unable to "test" his proposition against the archaeological record because his traits are not empirical in that context. The point is simple. Historical information was not used to create the generalization; it is not a generalization about history as is implied. The "sequence" is generated by an assumption of progress registered as the number of traits found in combination; it is not a set of societies arrayed in chronological order. Nowhere does Carneiro attempt to explain *why* Culture becomes more complex over time. Although he is circumspect in the use of the word "progress," his use of terms like "low-level traits," "advanced," and "primitive," reveals the pervasive role of the notion.

Carneiro does not link his observations about traits to the stages of the new Cultural Evolution explicitly. He does, however, supply a rationale for stages. He notes that the evolutionary distances between traits are not random in his scalogram and generalizes his observations as "in terms of the *time* required ... degree of difficulty *in evolving* a trait is inversely proportional to the rank order" of the trait (Carneiro 1968, p. 371, emphasis added). Thus, "because it does take *longer* for a simple *society ... to evolve* the next trait *due* in the sequence, at any given time a good many societies may be found at *the same phase of development*" (Carneiro 1968, p. 372, emphasis added). In addition to accounting for stages as a consequence of the difficulty of adding traits, these statements show how subtly number of traits becomes time and that the causative agent is assumed to be the abilities and/or desires of societies. While the pattern with which Carneiro is concerned is a consequence of his assumptions, the number of cases exhibiting particular combinations of traits, interpreted as rate of change, is a function of his sample. Carneiro recognizes this possibility and allows that at least some of the pattern may be due to his selection of cases. However, he quickly ties the pattern to one of the "facts" of cultural evolution, namely, that there is an "acceleration of evolutionary advance with increased cultural complexity" (Carneiro 1968, p. 372), and traces the notion that the rate of cultural change must proceed in geometric progression to Spencer (1880, pp. 446, 462). This geometric rate of change is one of the "facts" which I address later. Thus, in Cultural Evolution progress is its own cause. The notion of progress is used to create the sequence which is the "evidence" of progress.

One further aspect of the twentieth century version of Cultural Evolution merits attention. White's main concern had been with the evolution of Culture rather than cultures. In this he had been faithful to the generality of the Spencerian approach. But this unilinear chain of advancement implied the transformation of humanity, albeit apparently at different rates, through a unity of forms. The diversity of human cultures was completely unattended beyond complexity. While a reduction in diversity was plainly a feature of the late nineteenth and twentieth century human world, everyone acknowledged that this was linked to the expansion of Western culture at the expense, often catastrophically, of other cultures. The long term trend had plainly been in the opposite direction.

Partly in response to this need to account for diversification and partly because of the challenge presented by the ecologically oriented "multilinear" evolution of Julian Steward (1955), the new Cultural Evolution quickly came to encompass two evolutions. The one thus far described, with roots in the nineteenth century was termed "general evolution." It was contrasted with "specific evolution" (Sahlins 1960), which was taken to represent local ecological adjustment and specialization and seen as acting counter to the trends of general evolution. For example, a specific rule was formulated in which specialization was taken to be inversely related to the ability of a given culture to move from one stage to another (Service 1960). The Cultural Evolutionists saw specific evolution as analogous to biological evolution. This accommodation to diversity and the world of science was little more than accommodation. The mechanism of specific evolution is just as vague and vitalistic as that of general evolution. People, or rather societies, through intelligence, wisdom, foresight, or desire, adapt to local conditions, just as through the same agencies societies evolve from simple to complex. Selection was not part of specific evolution any more than it was a part of general evolution. In *Evolution and Culture* (Sahlins and Service 1960), the volume in which this dichotomy was first drawn, the word selection only appears twice, both times as a synonym for adaptation. In spite of the supposed contrast with general evolution, specific evolution retains the main methodological features of general evolution.

By the early 1970s it was apparent that the new Cultural Evolution had not revolutionized sociocultural anthropology in the manner White had envisioned. It was largely confined to the University of Michigan and students trained there, although because of the stature of the University of Michigan in anthropology, this did not relegate the new Cultural Evolution to insignificance. Segraves (1974) explains the new Cultural Evolution's apparent lack of broad appeal as a consequence of what she interprets as poor explication of specific evolution and an overemphasis on general evolution. Her attempt to clarify specific evolution amounts to defining it as progress in terms of thermodynamic efficiency in contrast to general evolution in which progress is measured by total energy captured. This reworking of specific evolution makes it even more similar to general evolution methodologically; it is the same program effected by a different kind of progress. Progress as efficiency runs afoul of the arguments advanced by Gould (1986) against optimality and those by Lewontin (1979) against adaptationist programs. This idea of progress has modest empirical basis. For example, Rappaport (1984) argues that while industrial societies capture more energy than other societies, they are not very efficient in energy use. Kelly's (1985) discussion of the Nuer and Dinka in East Africa demonstrates the expansion of a less efficient group at the expense of a more efficient group over a long period of time. Because specialization is equated with efficiency of energy use which is seen as progress in specific evolution, Segraves is compelled to regard contemporary culture as generalized in spite of the fact that it is totally dependent on a handful of plants, and even a handful of strains of plants (Rindos 1984).

Segraves's discussion is the last real attempt to develop Cultural Evolution as a methodology. If anything, the new Cultural Evolution is even less a force in anthropology today than it was in 1974 (Stocking 1987). Were it not for the fact that American archaeologists adopted elements of the new Cultural Evolution, Cultural Evolution would not be of further concern. That the source of these elements lies in American anthropology rather than European archaeology is clear. European Cultural Evolution in the hands of Childe was very much based in technological progress, not in sociocultural complexity. It is stages, defined in terms of sociocultural complexity, that are borrowed by archaeologists. The initial and still widest use of concepts derived from the new Cultural Evolution by archaeologists is in the study of the "origins of complex society" (Flannery 1972) where the stages Chiefdom and State (Service 1962) abut. In the main, the use of Cultural Evolution by archaeologists is benign; the major effort is to show that some particular set of archaeological data have or do not have some attributes that are surrogates for definition of a particular stage (e.g., Yerkes 1983). These associations then serve as the basis for reconstructing past societies, but since nothing is asked or expected of the reconstructions, no real harm is done. Cultural Evolution is not generally used to explain the course of history, though perhaps *sub rosa* it is assumed to. In spite of assessments like that of Coe (1977, p. 25) that " ... this has not gotten us very far in New World archaeology," the use of stages borrowed from the new Cultural Evolution seems to be spreading, restructuring the way in which research is conceived and the results integrated (e.g., Braun and Plog 1982). While the main thrust of archaeological usage has been directed at general evolution, recently Flannery and Marcus (1983) have focused attention on specific evolution, conceived broadly as an adaptationist history rather than thermodynamic efficiency. While this is certainly a more effective descriptive tool than a set of stages, it remains, to the extent that it purports to explain, a vitalistic nonselectionist approach like its parent. The actors are "cultures," the agents are the desires and wisdom of people in adapting, moderated by a history of isolation and contact.

In sum, while evolution means and has meant a variety of different things within anthropology and archaeology over the past century, they share a common structure. The underlying metaphysic is essentialist; causation is vitalistic, specifically attributed to the subject of study; and evolution is seen as progress, though progress itself has been given variable interpretation. Cultural Evolution is not related to evolution in the biological sciences, something that the better scholars have recognized, the more chauvinist have touted, but the majority have forgotten.

The Contrasting Roles of Progress

As this volume (and Huxley 1962; Vrba and Gould 1986) implies, the notion of progress is troublesome even within biology. The simple passage of time, because time is directional (Morris 1984), imparts directionality to all contained within it that changes. The linkage between forms and time effected by ancestor-descendant trait transmission necessarily limits the range of possible forms so that when looking back in time it is possible to discern lines of descent. Consequently, the simple presence of directionality or even temporally correlated trends and tendencies hardly warrants much attention in and of itself. Progress, thus it would seem, must attend some additional judgments about direction, trend, and tendency, i.e., judgments of higher and lower or better or worse.

From Darwin on biological evolution has attempted to avoid linking evolutionary theory to progress. However, the strong notion of progress that characterizes Western thought, and as objectified in Cultural Evolution, almost guarantees that progress in the sense of betterment will creep into any historical inquiry unless explicit theoretical measures are taken to exclude it. Bakker's (1986) analysis of the status of the Dinosauria seems to point to such occurrence. A notion of progressive relations among living representatives of animal classes structures the conception of the paleontological record through taxonomy, in such a way that the record conforms to the a priori substantive conclusion reached on the basis of ahistorical analysis. Yet to an outsider, and I am certain that in regard to matters biological, it seems that when progress is seriously and explicitly considered, it is usually a *description* of the course of evolutionary change, generalizations about trends and directions extracted from a history constructed independently of those trends and directions.

As Popper (1963) pointed out long ago, the status of such generalizations is suspect. Unlike the ahistorical sciences, which do not employ ratio scale time and in which generalizations may serve to stimulate profitable inquiry (e.g., as Keppler's "laws" of planetary motion became a subject of explanation for Newton), the asymmetry introduced by ratio scale time may reduce all substantive trends and generalizations in that dimension to accidental status. Bakker's contentions about the Dinosauria certainly point in that direction as does the role that Stanley (1987) accords mass extinctions in shaping the history of life. It is hardly my place to suggest whether biologists ought to regard what they may perceive as trends, tendencies and directions in the course of evolutionary change as suitable subjects for investigation. The point is, however, that apart from errors, progress is an observation about the courses of change.

Progress in anthropology and archaeology is also used in this sense. When data are arrayed in historical order, trends, tendencies, and directions of variable generality are detectable. If one wants to call such generalizations "progress," is aware of the status of such observations and labeling them progress does not preclude their explanation, there is little with which to disagree. Unfortunately, this is not the only use of progress and the second use is not as benign. In Cultural Evolution, progress also assumes the role of a theoretical tenet. This in turn structures the conception and interpretation of phenomena in such a way that progress is its own cause. While this was largely *sub rosa* in the nineteenth century version where archaeological data and their stratigraphic orders influenced the shape of progress, the new Cultural Evolution consciously elevated progress, in guise of "laws" of history, to theoretical status. In this role progress, usually seen as an inherent drive for increasing sociocultural complexity, explicitly and purposely structures data description as an interpretive algorithm.

Darwinian Evolution and Cultural Change

The most remarkable part of this account is that, in spite of explicit pretensions to science, open advocacy of evolutionary approaches, and insightful critiques of Cultural Evolution, Cultural Evolution is still the evolution of anthropology, with only a handful of exceptions. Certainly, the poor scholarship that led second generation evolutionists to confuse biological and Cultural Evolution, has, as Blute (1979) argues, played a significant role. More importantly, the largely atheoretical character of anthropology and archaeology, like other social sciences, renders them extremely easy targets for penetration by contemporary common sense (Dunnell 1982).

This is evident in the criticisms voiced of the few attempts to employ Darwinian evolution in the realm of cultural change. Rindos's (1980, 1984) account of the adoption and spread of agriculture, for example, has been criticized, not because it fails to explain or is contradicted by any physical evidence, but because it omits human intention as cause (Aschmann 1980; Ceci 1980; Schaffer 1980). Anthropologists face a problem not faced by other scientists -- their subject matter speaks to them. Indeed, what the subject says is at least as important a source of data as are observations of the subject. To complicate things further, many of the things people say are *reasons* which they use to account for their own actions. It is easy to appreciate how anthropologists and archaeologists frequently confuse these reason-giving motivational statements with scientific cause. Explanatory systems that rely on mechanisms and attribute cause to a theoretical system must, in their view, be a priori rejected or amended to include reason-giving in a causative role. Again, this is a fairly subtle matter because investigators have access to reason-giving only when the subject matter talks. This assertion also compels one to regard people as qualitatively different than other living things and serves to warrant a separate and different Cultural Evolution.

Working against the serious consideration of a Darwinian explanation of human history is the simple failure to recognize that a new theory requires new data. The essentialist underpinnings of Cultural Evolution manifest in the choice and description of archaeological and ethnographic data act to suppress almost all variation, the very information that is required to implement a selectionist model like Darwinian evolution. Extant data are congruent with Cultural Evolution because Cultural Evolution manufactured them. Without a major effort to acquire new data on variation, only selectionist just-so-stories can be erected on extant data, unsatisfying to both their authors and critics.

All of these elements combine to generate the feeling that use of Darwinian evolution is no more than a "tiresome biological analogy" (Flannery 1983, p. 1), incapable of explaining the "facts" of human history. While I am not exempt from the liabilities imposed by extant data, I think it is instructive to look at two of the "facts" of human history that are central to Cultural Evolution -- the ever increasing rate of cultural change over time and the contrast between simple and complex society.

Rates of Change

From Spencer (1880) on anthropologists have made much of the apparent geometric increase in the rate of change that has characterized, on the average, recent human culture. Inasmuch as change is caused by the people themselves in Cultural Evolution, the cliche that "necessity is the mother of invention" characterizes the motor running innovation. People desire to improve their lives or at least must solve problems presented by the physical world and it is this process that drives the rate of change. Further, human knowledge is cumulative, so the process is additional. Although human knowledge is certainly transmitted from generation to generation, it is silly to characterize all human knowledge as cumulative. Were it so, all traits would be scalable in Carneiro's sense. Archaeologists would not have to spend a large fraction of their time attempting to ascertain how particular kinds of artifacts were made and used. But the general nature of the trend itself cannot be denied. For the first million and a half years of human culture, a single set of tools was in use. The next couple of hundred thousand years are similarly characterized by another, only slightly more complex technology. This pattern continued in decreasing units of time to the modern era. How can a Darwinian evolution account for such obvious "progress"?

Elsewhere I (Dunnell 1980) have suggested that another similar trend may well hold the answer: the size of human populations. Human population is usually depicted as growing at a geometric or logarithmic rate (Deevy 1960). In itself such a pattern of population growth has interesting implications for the evolution of human beings (Dunnell 1980), but if it is true, even only approximately, the geometric or logarithmic increase in the rate of cultural change is an expected consequence in a Darwinian framework. If innovations, the cultural equivalents of mutations, occur once every so many man-hours on the average, simply increasing the number of people will have the effect of increasing the rate of cultural change proportionately.

Unfortunately, neither the rate of change nor the size of population is measured with any precision. Consequently, it is not possible to ascertain whether such a simple relation obtains between population size and rate of change, let alone analyze any residual differences that might indicate other factors affecting rate of change. Simon (1980), however, presents historical information on both parameters for Greek and Roman empires which suggests that population size does, even in complex societies, play a dominant role in determining rates of cultural change. Innovation rates, rather than peaking at times of stress as implied in a "necessity is the mother of invention" scenario, map the changes in population, both up and down, rather closely in the two cases. At least to the extent that the two parameters are understood today, there is no compelling reason to call upon human intention, desire, or even intelligence, to explain a dramatically accelerating rate of cultural change. It may be largely a simple, predictable function of population size.

Rise of Complex Society

Anthropologists have long recognized a fundamental division among human societies usually objectified as a contrast between simple and complex society. This contrast serves to distinguish states or stratified societies from the other stages in Cultural Evolution. Simple societies are those which are organized by relations of kinship, and they are constituted by a set of functionally redundant (i.e., excluding sex/age, all individuals perform all behaviors) individuals. In complex society, kinship plays no role in structuring the society; in fact, kinship is often seen as antithetical to the functioning of complex societies (Apter 1963). Further, members of complex societies are functionally differentiated in such a way that several kinds of individuals are required for the society to persist, either as a whole or as a collection of individuals.

As is the case with rates of change, there is an empirical basis for the simple/complex dichotomy and it requires explanation. Within the Cultural Evolution model a variety of different accounts for the rise of complex society have been proposed including the accumulation and control of surplus production (itself a controversial notion; Harris 1959; Orans 1966) by particular individuals, defensive responses to warfare, and the need to increase production in response to uncontrolled population growth (see Wenke 1981 for a review of explanations of social complexity). As various as they are, all of these accounts see societal complexity arising as a direct or indirect consequence of human attempts and desires rather than as a result of the operation of some mechanistic cause.

Anthropologists did not, at least initially, define social complexity or civilization using the criteria just outlined. Rather these criteria are an extensional definition of complex society extracted from the attributes of those units grouped as complex societies on other grounds. Defining social complexity in terms of functional redundancy and the absence of kinship as an organizing principle does, however, allow one to phrase the distinction between simple and complex, uncivilized and civilized, in terms that are meaningful in a Darwinian framework and thus to at least ascertain the nature of the distinction if not explain why and how it comes about.

As Wenke and I (Dunnell 1978, 1980; Dunnell and Wenke 1979) have suggested, when the distinction between simple and complex societies is phrased in this fashion, the difference between the two kinds of societies appears to be one of the scale at which natural selection is most effective. Simple human societies are organized like those of other social animals. Individual people carry the full code necessary to generate the full human phenotype, i.e., they are the units of reproduction and are functionally complete (Lewontin 1970). Further, simple societies are classic viscous populations (Hamilton 1964), groups of real or fictive relatives, living together. Complex societies, on the other hand, are genetically heterogeneous groups of people bound together by obligatory interaction. No one individual carries the complete "code" to reproduce the full phenotype. In short, the "individual" is an entity comprising many human organisms; the reproductive unit as well as the functional unit is a collection of people at least roughly equivalent to an anthropological society. Other features of a complex society, such as centralized decision making, are difficult to see as enhancing fitness on any scale if complex societies are simply groups of competing individuals. The very same features are expected if the society is the reproductive unit.

Such a shift in the scale at which selection is most effective is not without precedent. Gould (1977), for example, suggests that a shift from selection acting on individual cells to selection acting on groups of functionally differentiated and interacting cells may lie at the root of the Cambrian explosion of life forms. More recently Vrba and Gould (1986) describe such changes in terms of hierarchy and the bonding of lower-level individuals into higher-level entities. They suggest that if there is a weak but persistent vector that can be called progress, it is to be understood as a kind of "structural ratchet" (Vrba and Gould 1986, p. 226) pointing in the direction of increased complexity. It is interesting to note that Vrba and Gould (1986, p. 226) speculate on the possibility of lower-level individuals regaining their independence because the historical instability of complex society suggests precisely this kind of reversibility (Dunnell and Wenke 1980).

Identifying the nature of the complex/simple society dichotomy in an evolutionary framework is a necessary first step but is not an explanation of why complex society arises when and where it does. Culture, as a mechanism in addition to genetics for trait transmission, opened the possibility of larger units of selection by decoupling the transmission of traits from individual people. Rindos (1986) has offered a plausible account of how the capacity for cultural transmission of most behavioral parts of the human phenotype could have been fixed by selection, driving those genes which control specific behaviors to neutrality. It seems likely that substantial populations and a complex array of behavioral traits are both required before a shift in the size of the reproductive unit can take place. This is consistent with the relatively late appearance of cultural complexity in the human lineage. The specific conditions that drive the change are still obscure, but having conceptualized the problem in these terms there is no a priori reason why they cannot be isolated.

Conclusions

It is interesting to note that anthropologists have always implicitly treated cultures and societies as the units of change (see Marks and Staski n.d. for an explicit argument of this sort). To biologists, they were group selectionists. Given the atheoretical nature of anthropology and archaeology and the consequent deep penetration of their paradigms by contemporary common sense, the proclivity to treat cultures as the units of change is wholly understandable. Intuitively, cultures do compete, change and become extinct. But if my analysis of the nature of complex society is even approximately correct, Cultural Evolutionists have attributed a characteristic of complex society to all of humankind and used it to warrant a separate Cultural Evolution. To the extent that progress is an inherent property of organic change, it is a quantum phenomenon structurally ordained in shifts in the scale of selection. That such a change seems to have occurred recently in human history may go a long way to explain our fascination with progress.

Acknowledgments

I would like to thank K. Nyrop for bibliographic assistance and M. D. Dunnell for her editorial help. D. K. Grayson, P. V. Kirch, and M. H. Nitecki read the manuscript in draft and both the sense and the style have improved in consequence.

References

Apter, D. 1963. *Ghana in transition*. 2nd ed. Princeton: Princeton University Press.

Aschmann, N. 1980. Comment on *Symbiosis, instability and the origin and spread of agriculture: A new model*, by D. J. Rindos. *Current Anthropology* 21:765.

Bakker, R. T. 1986. *The dinosaur heresies*. New York: William Morrow.

Bayard, D. T. 1969. Science, theory and reality in "New Archaeology." *American Antiquity* 34:376-84.

Binford, L. R. 1968. Archeological perspectives. In *New perspectives in archeology*, ed. S. R. Binford and L. R. Binford, 5-32. Chicago: Aldine.

Binford, L. R. 1983. *In pursuit of the past: Decoding the archaeological record*. London: Thames and Hudson.

Blute, M. 1979. Sociocultural evolutionism: An untried theory. *Behavioral Science* 24:46-59.

Boas, F. 1896. The limitations of the comparative method of anthropology. *Science* 4:901-08.

Boas, F. 1911. *The mind of primitive man*. New York: Macmillan Co.

Boyd, R., and P. J. Richerson. 1985. *Culture and the evolutionary process*. Chicago: University of Chicago Press.

Braun, D. P., and S. Plog. 1982. Evolution of "tribal" social networks: Theory and prehistoric North American evidence. *American Antiquity* 47:504-25.

Carneiro, R. L. 1967. Introduction. In *The evolution of society: Selections from Herbert Spencer's principles of sociology*, ed. R. L. Carneiro, ix-lvii. Chicago: University of Chicago Press.

Carneiro, R. L. 1968. Ascertaining, testing, and interpreting sequences of cultural development. *Southwestern Journal of Anthropology* 24:354-74.

Carneiro, R. L. 1970. A quantitative law in anthropology. *American Antiquity* 35:492-94.

Carneiro, R. L. 1972. The devolution of evolution. *Social Biology* 19:248-58.

Carneiro, R. L., and S. F. Tobias. 1963. The application of scale analysis to the study of cultural evolution. *Transactions of the New York Academy of Sciences*, Series II, 26:196-207.

Cavalli-Sforza, L. L., and N. W. Feldman. 1981. *Cultural transmission and evolution: A quantitative approach*. Princeton: Princeton University Press.

Ceci, L. 1980. Comment on *Symbiosis, instability, and the origin and spread of agriculture: A new model*, by D. J. Rindos. *Current Anthropology* 21:766.

Childe, V. G. 1935. Changing methods and aims in prehistory. *Proceedings of the Prehistoric Society* 1:1-15.

Childe, V. G. 1946. *What happened in history*. New York: Pelican Books.

Coe, M. D. 1977. Archaeology today: The new world. In *New perspectives in Canadian archaeology*, ed. A. G. McKay, 23-38. Ottawa: The Royal Society of Canada.

Daniel, G. E. 1962. *The idea of prehistory*. Baltimore: Penguin Books.

Deevy, E. S. 1960. The human population. *Scientific American* 229(9): 195-204.

Dunnell, R. C. 1978. Natural selection, scale, and cultural evolution: Some preliminary considerations. Paper presented at the 77th Annual Meeting of the American Anthropological Association, November 14-18, Los Angeles.

Dunnell, R. C. 1980. Evolutionary theory and archaeology. *Advances in Archaeological Method and Theory* 3:35-99.

Dunnell, R. C. 1982. Science, social science, and common sense: The agonizing dilemma of modern archaeology. *Journal of Anthropological Research* 38:1-25.

Dunnell, R. C. In Press. Aspects of the application of evolutionary theory in archaeology. In *Archaeological thought in America*, ed. C. C. Lamberg-Karlofsky and P. L. Kohl. Cambridge: Cambridge University Press.

Dunnell, R. C., and R. J. Wenke. 1979. An evolutionary model of the development of complex society. Paper read at the 1979 Annual Meeting of the American Association for the Advancement of Science, San Francisco.

Dunnell, R. C., and R. J. Wenke. 1980. Cultural and scientific evolution: Some comments on *The decline and rise of Mesopotamian civilization*, by N. Yoffee. *American Antiquity* 45:605-9.

Flannery, K. V. 1972. The cultural evolution of civilizations. *Annual Review of Ecology and Systematics* 3:399-426.

Flannery, K. V. 1983. Theoretical framework. In *The cloud people: Divergent evolution of the Zapotec and Mixtec civilizations*, ed. K. V. Flannery and J. Marcus, 1-4. New York: Academic Press.

Flannery, K. V., and J. Marcus, eds. 1983. *The cloud people: Divergent evolution of the Zapotec and Mixtec civilizations*. New York: Academic Press.

Freeman, L. G. 1968. A theoretical framework for interpreting archaeological materials. In *Man the hunter*, ed. R. B. Lee and I. DeVore, 262-67. Chicago: Aldine.

Fried, M. 1967. *The evolution of political society*. New York: Random House.

Gould, S. J. 1977. *Ever since Darwin: Reflections in natural history*. New York: Norton.

Gould, S. J. 1986. Evolution and the triumph of homology, or why history matters. *American Scientist* 74:60-69.

Grayson, D. K. 1983. *The establishment of human antiquity*. New York: Academic Press.

Hamilton, W. D. 1964. The genetical theory of social behavior. *Journal of Theoretical Biology* 7:1-52.

Harris, M. 1959. The economy has no surplus. *American Anthropologist* 61:185-99.

Harris, M. 1968. *The rise of anthropological theory*. New York: T. Y. Crowell.

Hull, D. L. 1965. The effect of essentialism on taxonomy: 2000 years of stasis. *British Journal for the Philosophy of Science* 15:314-16; 16:1-18.

Huxley, J. S. 1962. Higher and lower organization in evolution. *Journal of the Royal College of Surgeons, Edinburgh* 7:63-179.

Huxley, T. H. 1863. *Evidence as to man's place in nature*. London: Williams and Norgate.

Kelly, R. 1985. *The Nuer conquest: The structure and development of an expansionist system*. Ann Arbor: University of Michigan Press.

Kuhn, T. 1962. *The structure of scientific revolution*. Chicago: University of Chicago Press.

Lane Fox, A. H. 1875a. Principles of classification. *Journal of the Anthropological Institute* 4:293-308.

Lane Fox, A. H. 1875b. On the evolution of culture. *Proceedings of the Royal Institution of Great Britain* 6:496-520.

Leonard, R. D., and G. T. Jones. 1987. Elements of an inclusive evolutionary model for archaeology. *Journal of Anthropological Archaeology* 6:199-219.

Lewontin, R. C. 1970. The units of selection. *Annual Review of Ecology and Systematics* 1:1-18.

Lewontin, R. C. 1974. Darwin and Mendel -- The materialist revolution. In *The heritage of Copernicus: Theories pleasing to the mind*, ed. J. Neyman, 166-83. Cambridge: MIT Press.

Lewontin, R. C. 1979. Sociobiology as an adaptationist program. *Behavioral Science* 24:5-14.

Lubbock, J. 1865. *Pre-historic times, as illustrated by ancient remains and the manners and customs of modern savages*. London: Williams and Norgate.

Lubbock, J. 1870. *The origin of civilization and the primitive condition of man: Mental and social condition of savages*. London: Longmans, Green.

Lyell, C. 1863. The antiquity of man. *Athenaeum* 41:523-25.

MacWhite, E. 1956. On the interpretation of archaeological evidence in historical and sociological terms. *American Anthropologist* 58:3-25.

Marks, J., and E. Staski. N.d. Individuals and the evolution of biological and cultural systems. *American Anthropologist*. In press.

Mayr, E. 1959. Typological versus population thinking. In *Evolution and anthropology: A centennial appraisal*, ed. B. Meggers, 409-12. Washington, D. C.: Washington Anthropological Society.

Mayr, E. 1982. *The growth of biological thought*. Cambridge: Harvard University Press.

McLennan, J. F. 1865. *Primitive marriage*. Edinburgh: Adam and Charles Black.

Morgan, L. H. 1877. *Ancient society*. New York: World Publishing.

Morris, R. 1984. *Time's arrows: Scientific attitudes toward time*. New York: Simon and Schuster.

Mulvaney, D. J. 1977. Classification and typology in Australia. In *Stone tools as cultural markers: Change, evolution, and complexity*, ed. R. V. S. Wright, 263-68. Canberra: Australian Institute of Aboriginal Studies.

Murdoch, G. P. 1967. *Ethnographic atlas*. Pittsburgh: University of Pittsburgh Press.

Orans, M. 1966. Surplus. *Human Organization* 25:24-32.

Popper, K. 1963. *The poverty of historicism*. Rev. ed. London: Routledge and Kegan Paul.

Rappaport, R. 1984. *Pigs for the ancestors: Ritual in the life of a New Guinea people*. Rev. ed. New Haven: Yale University Press.

Rindos, D. J. 1980. Symbiosis, instability, and the origin and spread of agriculture: A new model. *Current Anthropology* 21:751-72.

Rindos, D. J. 1984. *The origins of agriculture*. Orlando, FL: Academic Press.

Rindos, D. J. 1986. The genetics of cultural anthropology: Toward a genetic model for the origin of the capacity for culture. *Journal of Anthropological Archaeology* 5:1-38.

Sahlins, M. D. 1960. Evolution: Specific and general. In *Evolution and culture*, ed. M. D. Sahlins and E. R. Service, 12-44. Ann Arbor: University of Michigan Press.

Sahlins, M. D. 1972. *Stone age economics*. Chicago: Aldine.

Sahlins, M. D., and E. R. Service, eds. 1960. *Evolution and culture*. Ann Arbor: University of Michigan Press.

Schaffer, J. G. 1980. Comment on *Symbiosis, instability and the origin and spread of agriculture: a new model*, by D. J. Rindos. *Current Anthropology* 21:768.

Segraves, B. A. 1974. Ecological generalization and structural transformation of sociocultural systems. *American Anthropologist* 76:530-52.

Service, E. R. 1960. The law of evolutionary potential. In *Evolution and culture*, ed. M. D. Sahlins and E. R. Service, 93-122. Ann Arbor: University of Michigan Press.

Service, E. R. 1962. *Primitive social organization*. New York: Random House.

Service, E. R. 1975. *Origins of the state and civilization: The process of cultural evolution*. New York: Norton.

Simon, J. L. 1980. Bad news: Is it true? *Science* 210:1305-08.

Slotkin, J. S. 1952. Some basic methodological problems in prehistory. *Southwestern Journal of Anthropology* 8:442-43.

Sober, E. 1980. Evolution, population thinking, and essentialism. *Philosophy of Science* 47:350-83.

Spencer, H. 1880. *First principles*. 4th ed. New York: D. Appleton and Co.

Stanley, S. M. 1987. *Extinction*. New York: Scientific American Books.

Steward, J. H. 1955. *Theory of culture change: The methodology of multilinear evolution*. Urbana: University of Illinois Press.

Stocking, G. W. 1987. *Victorian anthropology*. New York: Free Press.

Tylor, E. B. 1865. *Researches into the early history of mankind and the development of civilization*. London: J. Murray.

Tylor, E. B. 1871. *Primitive culture: Researches into the development of mythology, philosophy, religion, language, art and custom*. London: J. Murray.

Tylor, E. B. 1881. *Anthropology: An introduction to the study of man and civilization*. New York: D. Appleton and Co.

Vrba, E. S., and S. J. Gould. 1986. The hierarchical expansion of sorting and selection: Sorting and selection cannot be equated. *Paleobiology* 12:217-28.

Watson, P. J., S. A. LeBlanc and C. L. Redman. 1971. *Explanation in archaeology*. New York: Columbia University Press.

Wenke, R. J. 1981. Explaining the evolution of cultural complexity; A review. *Advances in Archaeological Method and Theory* 4:78-128.

White, L. 1949. *The science of culture*. New York: Grove Press.

White, L. 1959. *The evolution of culture*. New York: McGraw-Hill.

Willey, G. R. 1961. Review of *Evolution and culture*, ed. M. D. Sahlins and E. R. Service. *American Antiquity* 26:441-43.

Willey, G. R. 1966. *An introduction to American archaeology*. Vol. 1. *North and Middle America*. Englewood Cliffs, NJ: Prentice-Hall.

Willey, G. R., and P. Phillips. 1958. *Method and theory in American archaeology*. Chicago: University of Chicago Press.

Wilson, D. 1851. *The archaeology and prehistoric annals of Scotland*. Edinburgh: Sutherland and Knox.

Yerkes, R. W. 1983. Microwear, microdrills, and Mississippian craft specialization. *American Antiquity* 48:499-518.

Yoffee, N. 1979. The decline and rise of Mesopotamian civilization. *American Antiquity* 44:3-35.

Morpho-Physiological Progress

Adam Urbanek

The present paper deals with the theory of morphological (or morpho-physiological) progress as formulated by the distinguished Russian comparative anatomist and evolutionary biologist Alexei Nikolaevich Severtsov (1866-1936). In the West his views are little known and almost never cited. However, Severtsov's ideas were extended and developed by many Soviet authors, and were exceptionally stimulating for at least two generations of biologists in the Soviet Union. What makes Severtsov's concept still important is his grasp and understanding of the dual nature of evolutionary changes. This duality can be understood as ambivalence, as complementarity, or as a dialectical unity of opposites. Whatever the approach, it is clear that Severtsov introduced a novel viewpoint, illuminating the nature of evolutionary processes. His elaborate terminology must not be mistaken for being merely much-ado-about-words. Severtsov used terms to denote important ideas and to emphasize the profound, qualitative differences among the evolutionary processes. Naturally, Severtsov's concept, formulated more than fifty years ago, needs critical evaluation and his descriptive-analytical categories should be redefined accordingly.

The Concept of Progress in Evolutionary Biology

Evolutionary progress belongs to those theoretical notions of biology that are never clearly defined, but which nevertheless play an important role in our understanding of the history of life. Philosophically, the concept of evolutionary progress presents a number of difficulties; it can hardly be defined and the criteria for its recognition are subject to an extreme diversity

195

of opinions (see Ayala 1974, and chapters in the present volume). In this respect it differs little from such fundamental biological concepts as homology, analogy, species, and evolution -- concepts also vaguely defined, but nevertheless theoretically operative.

The notion of evolutionary progress is closely related to a seemingly, or objectively, two-sided nature of evolutionary changes, namely adaptation and the level of organization. Such interpretation of evolution was clearly defined by Lamarck (1809), who identified a steady principle of progress (gradation) responsible for the universal and immanent rise of organization. While gradation is the main factor, adaptation is merely a secondary adjustment to the environment, a kind of evolutionary "noise," or the disturbance of the steady process of progress.

Darwin (1859) was less clear. Surely he considered adaptation as the primary, or major, effect of natural selection, but the changes or rise in the level of organization were considered a secondary, or cumulative, effect of numerous adaptive improvements in bodily structure and function. Moreover, Darwin was fully aware of the tremendous difficulties in recognizing or in measuring the level of organization of a given taxon, and he may be classified as an early relativist in respect to progress in animate nature.

Darwin postulated that natural selection is responsible for a continuous process of improvements and gradual rise in the organization level, while at the same time he emphasized that the action of natural selection does not imply any necessity of progressive development. He was probably unable to choose his criteria of progress in evolution: should they be adaptive (increased competitiveness, expansion) or organizational (complexity of structure, efficiency, etc.). Yablokov (1968) concluded that these questions constitute a real contradiction in the Darwinian theory, a contradiction resolved only by Severtsov's concept of biological and morpho-physiological progress.

Even earlier, Darwinism was frequently criticized for its failure to explain properly the adaptive improvements and the rise of organization. To Nägeli (1884) natural selection provided an explanation only for the change and perfection of adaptive traits, but did not explain the emergence and rise of new levels of organization. Why were liverworts not preserved forever as liverworts or worms as worms? Why were some liverworts transformed into cormophytes and some worms into arthropods? These are everlasting questions, largely immune to the growth of empirical knowledge.

The concept of progress provides a theoretical instrument for the analysis and understanding of the Janus-face of evolution, as revealed in the changes of adaptation as well as in organization. It is mostly a theory of the emergence of new levels of biological organization.

The ideas of morpho-physiological progress as advanced by Severtsov offer a certain solution to this burning problem of evolutionary biology. This deeply Darwinian solution is based on the diversity of the processes involved in evolutionary changes.

Severtsov's Classical Theory

Severtsov's (1931, 1939) concept of morpho-physiological progress constituted the core of the ideas of evolutionary progress as later developed by the Soviet biologists and philosophers. Severtsov considered evolutionary progress as an objective feature of long-lasting evolutionary changes. He argued against adaptive relativism, treating all organisms as equally well-adapted and thus rejecting any idea of progress, and at the same time he argued against concepts of general, immanent perfection, in which every change is by its very nature progressive.

According to Severtsov the weakness of earlier views on evolutionary progress lay in the lack of discrimination between *biological* and *morphological* aspects of evolutionary progress. His careful distinction between biological and morphological (or morpho-physiological) sides of evolutionary progress can be misleading to contemporary readers. However, he followed an established tradition of his time to differentiate biology (meaning mainly ecology and ethology of organisms) from morphology (meaning structural aspects of organization). Thus the biological side of progress refers to adaptive functions of evolutionary change, while the morphological aspect appeals to the complexity and efficiency of organization attained.

Due to control by natural selection the evolutionary processes are, at any given time, adaptive and therefore *always result in biological progress*. Biological progress not only means the survival of a particular species or higher group due to feeding and reproductive successes, but is especially characterized by (1) increase in number, (2) territorial expansion, and (3) taxonomic differentiation. Thus, biological progress is an inevitable result of the opportunistic nature of evolutionary change, produced by a number of different modes of phylogenetic processes. Severtsov distinguished four main modes of such changes, each ensuring biological progress but differing fundamentally in the nature of the processes involved. *Aromorphosis* occupies the crucial position among the four modes of phylogeny as it leads to biological progress as well as to improvements of morpho-physiological organization of the body (*morpho-physiological progress* according to Severtsov). To Severtsov aromorphosis meant (1) a phylogenetic change characterized by a general increase in "*vital energy*" and activity of organisms, and (2) a rise in intensity and efficiency of key organ functions. Severtsov convincingly demonstrated that evolutionary changes cannot be measured merely by the biological successes of populations, species or lineages. He suggested that evolution also involves an entirely new dimension -- an increase in general (versus specific) adaptation. Thus he distinguished two complementary aspects of evolutionary progress: *biological*, a partial improvement, and *morpho-physiological*, a general rise in the efficiency of organisms.

A general "increase in vital energy" by modern standards could hardly be considered an adequate definition, but its meaning becomes more understandable in the context of Severtsov's examples of aromorphosis. These examples are: (1) origin of external gills in lower vertebrates (as opposed to internal gill sacs of Agnatha), permitting better oxidation of blood; (2) transformation of one gill arch into jaw apparatus in Gnathostomata, which enables an active capture of prey (as opposed to passive intake of food by its forerunners), an adaptation so successful that it is preserved in all vertebrates including Man; and (3) the subdivision of the heart in terrestrial vertebrates, producing in birds and mammals a quadripartite heart that ensures the complete separation of arterial and venal blood, thus raising the oxygen level in body tissues. Such changes "improve" and intensify the functions of key organs, raise the efficiency of metabolism, and/or ensure increased activity. They are thus very broad and general adaptive gains, useful under most environmental conditions.

Moreover, the essential meaning of aromorphosis is compared with an alternative mode -- *idioadaptation*. In idioadaptation there is no increase in general efficiency or activity; it represents evolutionary changes leading to adaptation to specific environmental conditions and ensuring biological progress without changes of morpho-physiological organization. Idioadaptation is the prevailing mode of evolutionary change, while aromorphoses are relatively rare events. Specializations, according to Severtsov, are only one form of idioadaptation.

Two other modes of phylogenetic changes are less significant. *Degeneration* means (1) regression in the morpho-physiological organization, (2) a decrease in the efficiency and intensity of functions of the organs, (3) reduction of organs, or (4) simplification of the entire organization of the body. The development of parasitism and the change from a free to a sessile way of life are examples of degeneration. *Coenogenesis*, to Severtsov, represents embryonal adaptations of great significance for survival and for biological success, but with no effect on the adult stages of descendants.

Against the background of contemporary theories Severtsov's views represented a great advance toward a better understanding of certain theoretical notions, which until his time were only loosely defined. He succeeded in substantiating the composite nature of evolutionary progress, clearly distinguishing two aspects of the phenomena and defining the events leading to improvements of morpho-physiological organization as a distinct class of evolutionary change (aromorphosis). However, Severtsov's language, especially his understanding of aromorphosis as a "general rise of vital energy," created some problems. There were many attempts to redefine and modify the meaning of aromorphosis.

Schmalhausen's Contribution

Schmalhausen's *Routes and Modes of the Evolutionary Process* (1939) represented in the Soviet Union a turning point, because of elaboration of the classification of modes of evolution and extension of the meaning of aromorphosis to a previously unknown evolutionary role. The broad category of idioadaptation was subdivided into *allomorphosis* and *telomorphosis*.

The term allomorphosis describes those idioadaptations which paved the way for aromorphoses, went halfway toward them, or at least represented the direction of evolution not excluding eventual morpho-physiological progress. Telomorphosis describes extreme specializations which will never undergo an aromorphic change. This emphasized the duality of Severtsov's category of idioadaptation. However, Schmalhausen was bitterly opposed to adding more terms to the already ample nomenclature. In defense, Schmalhausen stressed the significance of this terminology which made better classification and understanding of the diversity of evolutionary processes.

Still more important was broadening the meaning of aromorphosis. Schmalhausen pointed out that aromorphosis allows for changing the environment and for better utilization of resources. Later Davitashvili (1961) called this the ecogenetic role of aromorphosis. According to Schmalhausen morpho-physiological progress implies a greater isolation from the direct influence of environment and, generally, a more active way of life.

But really crucial was the introduction of the idea of *epimorphosis* as unlimited evolutionary progress. By this term Schmalhausen meant a rise in organization and a widening of the environment that resulted in the conquest of the entire planet and exploitation of all its resources. This meant attainment of superiority over all other living beings and considerable control over the environment. In a given geological epoch epimorphosis logically could be realized by a single species only, and the emergence of Man is its only known example. Epimorphosis, in this case, involved development of the mind, speech, and culture as well as cumulative cultural transmission. These factors also enabled the development of human personality within the social milieu.

Epimorphosis, said Schmalhausen, provides above all, a graphic example of biological progress, best expressed in a rapid increase of human population, and its geographic expansion only later replaced by quite different factors of social development responsible for the Ascent of Man.

The Debate on Evolutionary Progress

Severtsov's ideas of evolutionary progress were an important intellectual legacy to the majority of Russian and Soviet biologists. Only a few Russian experimental biologists, especially the geneticists Filipchenko and Koltsov, were skeptical (see Adams 1980), and stressed the inadequacy of comparative and descriptive anatomical methods for the study of evolution. This criticism may be only partly ascribed to the well-known controversy between experimental biologists on one side and morphologists and paleontologists on the other. As we may now suppose, the attitude toward Severtsov and his work was to a large extent politically motivated.

This animosity stemmed from the 1911 unrest of the Moscow University students who defended the partial autonomy of the universities gained in the 1905-1907 revolutionary period. In protest against the police persecutions twenty-one professors (equivalent to Western department heads) and a large number of lecturers resigned. Among them were the distinguished zoologist M. A. Menzbir and the plant physiologist K. A. Timirazev. Moscow University was left almost without a teaching staff and ruling authorities made every effort to recruit new scientists for the emptied chairs. Severtsov, professor at Kiev University, was offered Menzbir's professorship. He accepted, probably with the conviction that science is above politics. However, Koltsov rejected the same offer with distaste. Severtsov's scientific merits and great teaching achievements did not excuse his decisions and actions, which were condemned and boycotted by his colleagues. Years passed before the animosities decreased somewhat, but he was never forgiven (Granin 1987).

Therefore extrascientific as well as scientific reasons shaped the contemporary attitude toward Severtsov's evolutionary concepts. Scientific criticism was to a great extent exaggerated. It was true that to understand evolution, morphological data needed completion through experiments. It was also true, for modern synthesis, that properly evaluated morphological information be fruitfully used.

Schmalhausen demonstrated how morphological and embryological studies and Severtsov's "speculative" concepts could be integrated with genetics and the theory of natural selection, and how they created the historical cornerstone for the emergence of the synthetic theory of evolution.

In spite of the firm belief of experimentalists that the entire field of comparative anatomy was dead, Severtsov's views proved to be attractive and stimulating for many of his followers and opponents. This was best expressed by the debate over his ideas of evolutionary progress in the following decades.

This debate revealed new aspects of the subject and has been summarized by Matveev (1967) and Davitashvili (1956, 1972, 1978), and especially in two chapters of a more recent fundamental monograph, *The Development of Theory of Evolution in the USSR* (Miklin and Mirzoyan 1983). They deal with the Severtsov concept as well as with the entire system of meanings concerning the evolutionary progress. A separate chapter by Yefimov (1983) is devoted to the evolutionary aspects of the origin of Man (anthropogenesis). In the postwar period no less than three distinct trends in the debate can be distinguished.

First, considerable controversies centered around attempts to generalize the meaning of aromorphosis. They represented tendencies to transform morphological understanding of aromorphosis into a more generalized formula that would include ecological or ecogenetic aspects of processes. Also, emphasis on the macroevolutionary significance of aromorphosis as well as an improved terminology (the use of "arogeny" instead of "aromorphosis") was proposed.

Second, Severtsov's ideas were further developed by stressing a profound analogy with technical and social progress. The concepts of progress in biological and social spheres inspired by Marxist dialectical and historical materialism were bound into a single theoretical system.

Third, but not least, there was advance in the elaboration of the epimorphosis concept. This idea of Schmalhausen's was treated as a major theoretical achievement crucial for the understanding of the origin of Man and his social and cultural advances. Here again the ideas of evolutionary progress were borrowed from dialectical and historical materialism.

The backbone of Soviet ideas on evolutionary progress has always been primarily biological. But it is also true that after the Revolution many naturalists have been influenced by Marxist theories of nature and society. The complex history of interactions between the philosophy of biology and Marxist philosophy had a number of turning points and bore positive and negative effects.

The predominance of materialistic philosophy in the post-Revolutionary period was responsible for sympathetic attitudes toward the idea of *development* of nature and society, and the idea of *progress* was usually understood as an important aspect of the development of the most complex forms of matter only. While development is a broad philosophical category meaning an ascendant or descendant directional change, progress was considered the main mode of development of complex teleonomic systems (Zavadsky 1970).

Although progress as a philosophical notion is less significant than development, this essentially positive attitude created a stimulating atmosphere for its study which was continuously scrutinized and treated as a fertile theoretical concept.

On the other hand, Marxist ideas of progress and retrogression developed chiefly in relation to social phenomena. Their direct application to biology was prevented by the understanding of various forms of motion as a qualitatively different and nonreducible phenomena. In this way studies of evolutionary progress had a relatively autonomous status, and were secure against the most simplistic and extreme views which were frequently propagated.

Controversies about Generalized Meanings of Aromorphosis

One of the notable trends in the further development of Severtsov's concept of aromorphosis was marked by attempts to modify, broaden, modernize and extrapolate the meaning of this term. Severtsov's views were almost a swan song of the heydays of evolutionary morphology and his theory coincided with the eclipse of comparative anatomy and phylogenetic approaches to the study of evolution. There is no surprise then, that his ideas soon appeared old-fashioned, especially since morphology was under strong fire from experimental biology (see above).

For these reasons attempts were made to introduce ecological criteria of aromorphosis and to avoid vaguely defined criteria by Severtsov (Schmalhausen 1939; Davitashvili 1956; Zavadsky 1953). Moreover, there were suggestions to replace aromorphosis by arogenesis, seemingly a better suited term for evolutionary biology (Tachtajan 1951; Zavadsky 1967). But this change implied also its different theoretical function. However, there was a reaction against suggested changes and a strong preference to retain the original meaning of the term for the sake of clarity in order to avoid corruption of its main theoretical notions (Tatarinov 1976). I will now detail the story of the debate.

Davitashvili, author of valuable studies on the history of evolutionary concepts in paleontology, was convinced that the idea of progress was crucial to the understanding of the major features of the history of life as revealed by the fossil record, but he was somewhat critical of Severtsov's views on evolutionary progress. This is especially true in his first book (Davitashvili 1956), written under the influences of Michurinian-Lysenkoist biology. In contrast to Zavadsky (1953), who considered Severtsov's theory as the best, Davitashvili stressed imperfections and weaknesses of the theory of aromorphosis. While Davitashvili accepted the fact that Severtsov rejected the heredity of acquired characters, he nevertheless emphasized a Darwinian component of Severtsov's philosophy. Davitashvili even defended Severtsov from the most ardent neophytes of Lysenkoism.

Davitashvili was critical of the meaning of Severtsov's "biological progress," as he believed it to be Severtsov's weakest point, and he considered the use of the term "progress" misleading in that context. What was actually meant was merely the survival and the flourishing of the groups. It was aromorphosis that fully deserved to be called evolutionary progress. Moreover, he rightly pointed out that Severtsov's description of aromorphosis as a "rise in the energy of life processes" may cause misunderstandings. It cannot be reduced to "vigor" (Davitashvili 1956), nor to the rate of metabolism (Davitashvili 1961). He tried to replace this vague description with an idea of ecological expansion which implies the widening of the adaptive zone and penetration of a new environment as an immediate result of aromorphosis. This idea gradually embraced the notion of ecogenesis, that is, the appearance of ecological novelties, the emergence of radically new ways of life, and the discovery of new habitats. The term "ecogenetic expansion" was coined (Davitashvili 1961) to describe what was believed to be the most important function of aromorphosis.

Fifteen years after his first book, Davitashvili (1972) published a monograph on evolutionary progress. He advanced a view that "evolution of the organic world does not represent disordered and random changes but a directional development, in principle, an ascending process from lower to higher." This, for him, was a fact distinctly outlined by the history of major groups of fossil animals. The main body of his monograph concerns the phylogeny or the "real routes of evolution in main groups of plants and animals." Davitashvili's book is an unsurpassed source of information on a broadly defined field of evolutionary progress. However, his theoretical considerations are in many ways unique, and are strongly emphasized in his *The Theory of Evolution* (1978), where he rejects what he calls the postneo-Darwinian viewpoints (= synthetic theory of evolution) and develops his own version of classical Darwinism. He disagreed with Yablokov (1968) who interpreted Darwin's position on evolutionary progress as indeterminate, and used Darwin to support his main theses. Previous criticism of some aspects of Severtsov's concepts were repeated, but in general the theory of aromorphosis as highly heuristic and valuable was accepted.

A rather interesting point is Davitashvili's changing attitude on his own principle of ecogenetic expansion; he once considered it as "the main feature of aromorphosis to which all remaining characters of evolutionary progress are subordinate" (Davitashvili 1978, p. 256), only later to present it as a contingent property of aromorphic changes. "Ecogenetic expansion is not always related to aromorphosis and the latter not always is the main cause of the former;" moreover, "evolutionary progress cannot be reduced merely to ecogenetic progress .... One cannot identify these two phenomena ..." (Davitashvili 1978, p. 365). It is impossible to decide whether these incongruencies reflect imperfection in Davitashvili's views or are merely a result of exceedingly rich and varied factual material, difficult to synthesize.

Another suggestion to modernize the classical meaning of aromorphosis came from Tachtajan (1951, 1954), a renowned Soviet botanist. He thought Severtsov's term should be generalized and deprived of its one-sided morphological implication. Therefore he replaced the term *aromorphosis* with *arogenesis*, which logically and linguistically better described any rise in the morphological, biochemical, and other organizations. This proposal was accepted by a number of authors (Paramonov 1967; Zavadsky 1967, 1971), and frequently treated merely as a formal correction. Usually both terms were considered synonymous, yet their usage shows a characteristic difference. Arogenesis has been commonly used in connection with the origin of higher categories. Zavadsky (1972) was inclined to consider aromorphosis as the main mechanism of macroevolution. He introduced the notion of arogenic populations and defined them as small populations with specific variation and direction of changes, capable of future macroevolutionary changes. Other Soviet authors even suggested that arogenesis is a general and universal way of the origin of higher taxa.

All these attempts to improve, modify or widen Severtsov's meaning of aromorphosis met with severe criticism by Tatarinov (1976), a well-known Soviet paleontologist. Application of ecological criteria reduced to some extent the one-sided morphological treatment of Severtsov and produced a somewhat illusory effect of a more profound biological attitude to aromorphosis. This caused a radical change in the sense of the term. Aromorphosis has been equated with the entire evolutionary progress (see especially views of Davitashvili discussed above), thus depriving it of the important point of Severtsov, namely that aromorphosis is only one of many possible modes of progress. Tatarinov's conclusion is clear: aromorphosis should be defined by morphological criteria, as originally proposed by Severtsov. As in the classical theory of Severtsov, to Tatarinov, physiological criteria play a secondary role only. Tatarinov was convinced that such redefinition will restore the primary sense of the term aromorphosis and will result in a greater clarity of meanings.

In this way the main criterion of aromorphosis should be the increase of the complexity of organization traced in particular evolutionary lines. Tatarinov argued that while a comparison of the complexity of organization of remote groups of different phyla is hardly possible, a conclusion within a single group on the progressive or nonprogressive nature of evolution can be drawn by comparison of representatives within this line. The criteria of progress should be independently selected in each case and thus would be different for bony fishes and for birds.

Tatarinov also objected to interpretation of arogenesis (redefined aromorphosis) as a universal mode of the origin of higher taxa, some of which do not represent any rise in the level of organization, and therefore cannot originate by means of arogeny (= aromorphosis). These brief considerations of Tatarinov also provide a general insight into questions of evolutionary progress.

In conclusion, it seems that searching for general evolutionary progress and its universal criteria might be futile, but looking for partial progress within particular lines or taxa is a meaningful practice. Aromorphosis is not an equivalent to macroevolution as developed in Western Europe and the United States. It pertains to an important fraction of macroevolutionary events, namely the origins of new Bauplans and breakthroughs to major adaptive zones. Thus there is a partial overlap between these two categories of evolutionary processes. However, aromorphosis estimates the effect of changes in terms of organization levels and efficiency, while macroevolution is usually defined by taxonomic effects. Moreover the theory of aromorphosis tells us nothing about the relation of microevolutionary processes to large-scale evolution, while in the theories of macroevolution the relation of *micro*- to *macro*- processes is essential. Thus aromorphosis and macroevolution are complementary, but not identical views on the diversity of evolutionary changes.

Evolutionary Versus Technological and Social Progress

Progress has been generally understood by Soviet philosophers as an important feature of the development of complex, material systems, and is seen in technology as well as in the development of the human society.

Theoretical problems of the progressive development of life and of technology were widely discussed in the symposium organized by the Institute of History of Natural Sciences and Technology of the Soviet Academy of Sciences in Leningrad in 1970. At this symposium Zavadsky (1970) formulated general ideas of the meaning of progress in living technical systems. Progress in living nature is in important aspects *isomorphic* to technological progress. There are essential similarities of patterns or regularities in both spheres. A progressive acceleration of changes, a trend toward increasing complexity, a growing autonomy (isolation from the outer noises), and reliability are believed to be among the common features. Miklin (1970) demonstrated an optimal relation among (1) the differentiation and the integration of structure and function, (2) the reliability of the system, (3) efficiency and energetic economy, (4) relative independence (isolation) of the system from outer influences, and especially important, (5) the quantity of information and quality of a control system.

However, it is clear that to all Soviet authors causation is fundamentally different in spite of similar patterns. Technology is goal-directed, man-dependent, and subject to social and economic control, while progress in living nature is the side effect of blind forces of natural selection controlled by stochastic processes. Therefore, the analogy between a man-made and living object is deeply misleading and must not be interpreted as a model of causal relations.

On the other hand, comparative studies of the theory of the progressive development of living and technical systems are among the most interesting approaches to a better understanding of biological evolution.

The biotechnological approach was first outlined by the German zoologist Franz (1935) who, in order to draw analogies, attempted to compare living organisms with technical devices. He coined the term, *biotechnical progress*, by which he meant the improvements of structure and function of organisms that can be measured by their efficiency. During progressive evolution the efficiency of the entire life cycle and of particular functions increases, and possibly such other parameters as profitability (Ertrag) and productivity (Leistung).

An improvement in the biotechnical efficiency of the organism may result in an adaptive advantage and evolutionary superiority of a given taxon. The evolutionary superiority is generally expressed in "the degree of elevation," which Franz characterizes as the approximation of the structures and functions of organisms to theoretical optimal solutions which define their competitive superiority.

Franz proved that biological and technical systems have common principles of organization, and that efficiency may be used to evaluate the degree of perfection of their organization (Kapralova and Lukina 1970). Biotechnical improvements of great functional significance may also constitute the essence of aromorphic changes.

The profound analogy between tools, machines and living organisms has been widely explored by modern technology (bionics). Practical use is the best evidence for the objective nature of this analogy. Thus the simulation of life processes, once an approach of a purely cognitive nature, has been transformed into a practical program of great economic significance. Therefore isomorphism, between organic and technological evolution, is used in bionics to design machines and instruments, while technical parameters are used to estimate the level of organization of living beings (Zavadsky 1970).

Zavadsky and Kolchinsky (1972, 1977) advanced a view that progress can also be seen in the basic mechanisms and factors of evolution, and they distinguish the *general* and the *partial* factors of evolutionary changes. Heredity and its variability (mutagenesis) plus natural selection provide an example of the general group of factors. Mutagenesis and natural selection represent the invariant causes of evolution, ever present since the prebiological stage was completed and replaced by biological evolution proper. However, these universal and invariant factors are accompanied by less common factors which are directly related to the level of organization of a given biosystem. This is especially true of the organization of its genetic apparatus (origin of chromosomes, diploidy, mitosis and meiosis, recombination, etc.) as well as in respect of the structure of its breeding systems (agamic, uni- and biparental, etc.). These factors also determine the course of important evolutionary processes.

The development of these partial factors and mechanisms of evolution undoubtedly represent a historical sequence of events, although their exact geochronology cannot be established. In other words, this is the problem of "the evolution of evolution."

Zavadsky and Kolchinsky (1972, 1977) were convinced that the predominance of certain factors and the replacement of some mechanisms by others allows one to distinguish certain *natural major stages* in the evolution of the living world. They noted an analogy with socioeconomic formations, as defined by Marxist theory, of historical materialism. Such formations represent historical types of society with a definite mode of production, while replacements of one formation by another constitute main events in the progressive development of the human history. Likewise an *evolutionary biological formation* represents a major stage in the evolution of the organic world, characterized by specific patterns of interaction with the general factors of evolution, and some partial factors unique to a given stage. This is an ambitious attempt to unify biological and social progress into a single theoretical system, although the qualitative difference between them is clearly recognized. Zavadsky and Kolchinsky (op. cit.) appeal instead to a remarkable similarity in the generalized pattern of historical and evolutionary processes, stressing at the same time their different causation and substrate.

An example of such a formation, discussed in detail by Zavadsky and Kolchinsky, is the Early Proterozoic formation with *substrate* or *biosystems* represented by prokaryotic organisms. The crucial features of their organization were the presence of a cell wall granting a certain degree of isolation from the direct influences of the environment, combined with the lack of a nuclear membrane as well as an indistinct differentiation of genotype from phenotype (phenotypic traits were very close to the genes!). Generation turnover was extremely rapid, elimination usually had a mass character, and population size was subject to drastic changes resulting in the large role played by genetic drift. This formation was a considerable advance over the eobiont or protobiont formation of the Archean, and was manifested, first of all, by a much greater reliability in the transmission of genetic information.

Zavadsky and Kolchinsky emphasized the tentative character of their concept, and the need of more studies to properly expose and elaborate the details. However, I can raise certain doubts immediately.

While successive socioeconomic patterns of human history represent an ascending order from lower to higher stages of development, as interpreted by historical materialism, there is no evidence that this is also the case with evolutionary-biological formations. It is, for example, quite probable that prokaryotes and eukaryotes evolved from a common ancestor and were living simultaneously during the Early Precambrian. We cannot consider these fundamentally different types of organizations as successive major stages in the history of life, but rather as alternative ways of solving certain evolutionary problems. Moreover, the genetic system of eukaryotes may be considered, in some respects, superior to that of prokaryotes, while in others (genes in pieces!) its organization seems even less effective. The hypothetical Archean "viral formation" hardly can be considered an independent stage in the evolution of life as suggested by Zavadsky and Kolchinsky. It seems more probable that viruses appeared and evolved as a part of an interacting bacteria-phage system. These and other questions indicate that the hypothesis of evolutionary-biological formations is probably premature, although we can consider certain stages in the development of mechanisms of evolution.

Evolutionary Process and Human Origin

Evolutionary changes which take place during *anthropogenesis* -- the process of the emergence of man -- can be characterized by a number of theoretical terms. The notion of unlimited progress, introduced by Huxley (1942), is one of these terms. While a vast majority of phylogenetic lines inevitably end with extinction or terminate with persistent forms, the line of man represents a notable exception. Man dominates the entire organic world and is potentially capable of further, and logically, unlimited progress. As Huxley succinctly put it ". . . progress hangs on but a single thread. That thread is the human germ-plasm . . ." (Huxley 1942, p. 572). Such broad prospects of the human evolutionary future stem from the unique characteristics of man, who transcended the biological sphere and entered the realm of psychosocial existence.

The Soviet authors believe that unlimited progress not only refers to the events which occurred during anthropogenesis or during primate evolution, but, as introduced by Huxley, unlimited progress extends into a long series of events preceding the final stages of human phylogeny and shapes further evolutionary advances in the direction of recent Man. Unlimited progress may be traced throughout the entire history of life from eobionts to modern human species (Zavadsky 1958). Therefore, unlimited progress is an evaluation of past events and the potential for future changes.

Schmalhausen's meaning of epimorphosis and the meaning of unlimited progress partly overlap; epimorphosis refers only to the final stages of aromorphic changes in primate evolution; while unlimited progress refers to a much longer sequence of events and it postulates man's exceptional abilities to survive and socially evolve.

Yefimov (1972) compared epimorphosis with Huxley's concept of unlimited progress and the classical aromorphosis of Severtsov. Epimorphosis represents only one possible type of aromorphosis, but it is a unique process, logically the last imaginable stage of biological evolution which crosses the purely biological laws. Producing a radically new adaptive type and distinct morphological changes, epimorphosis is clearly a macroevolutionary event. Due to the vast geographic expansion and population explosion, the final stages of human evolution may be considered as a mega-arogenesis. However, behavior, ethological factors, and superorganismal social systems of organization are even more important than biological success for understanding the very nature of epimorphosis (Yefimov 1972).

Yablokov (1968) analyzed the problems of unlimited biological progress but not the concept of epimorphosis. He was convinced that the meaning of unlimited progress (as opposed to the limited morpho-physiological progress) is a valid and important concept provided its criteria are properly defined. One of the criterion, namely the degree of approximation of man as the highest form of organization of matter, may be anthropomorphic. The alternative criterion, pertaining to the main stream of development, is free from this weakness. The meaning of the mainstream of evolution plays a significant role in the Soviet debates on evolutionary progress. Zavadsky (1958) proposed an even more elaborate classification, distinguishing three categories of progressive directions of evolution: the *principal mainstream*, leading to man; the *mainstream*, generally progressive but unrelated to the human line; and the *blind alleys*, side branches, sometimes flourishing, but without perspectives of further advance. These descriptive categories of evolution are very broadly and somewhat arbitrarily defined. Moreover, it is very difficult to assign actual evolutionary lines to any of the categories -- a point mercilessly criticized by Davitashvili (1972), who competed with Zavadsky for leadership in the field of evolutionary progress.

Davitashvili (1961, 1972, 1978) was critical of the idea of unlimited progress. A possible conclusion from Huxley's consideration, according to Davitashvili, is that evolutionary progress is rare or even exceptional while the available data indicates a contrary view. Moreover, division of the classification of progress into two clearly cut categories, "limited" and "unlimited," is inadequate and simplistic. Partial improvements, such as idioadaptations, also may potentially open new prospects for future evolution. Therefore, the subdivision of the objective processes into aromorphoses and idioadaptations, and even more into "limited" and "unlimited" progress, are to a large extent arbitrary procedures. Davitashvili was also skeptical that only a single line of phylogeny is the carrier of the "torch of progress." He even suspected vitalism or finalism in such a view on evolutionary progress.

Besides, Davitashvili's criticism of Huxley's approach to the exceptional role of man is vague. It is not clear whether future evolution will rely solely on the human species since it hinders the development of other higher forms of life. Nor could it be known, in the absence of man, if the living world would not develop progressively toward new psychosocial forms of existence. But, there is, of course, no reason to believe that the disappearance of man would block the "standard" evolution of other groups resulting in a new biological progress and emergence of new taxa.

Yefimov (1981), in the light of dialectic materialism, also attempted to evaluate the significance of Schmalhausen's epimorphosis in relation to the general theory of anthropogenesis. The concept of epimorphosis is the main evolutionary basis for understanding and explaining the transient ape/man period. For Yefimov, anthropogenesis, in broad philosophical terms, is a classical example of a sublation of the biological by the social. Sublation is a Hegelian notion meaning simultaneous cancellation and preservation, and is considered by Soviet Marxists to be an objective regularity in the development of the material world. Sublation is intimately related to the dialectic nature of change through "negation of negation" and stems immediately from it. As applied to the origin of Man, sublation is a certain mode of accumulation and incorporation of evolutionary novelties into the already existing system, accompanied by transformation as well as restructuring of the system itself in relation to the new whole. Sublation, so defined, presents an indispensable mechanism of passage from a biological to a social form of motion.

Paraphrased in the language of natural sciences, anthropogenesis, a dialectical process of change, may be described as a self elimination of Darwinian natural selection and a simultaneous increase of the role of biosocial selection. It is subsequently replaced with new factors related to social and labor activity. At the same time these changes are the main rules of epimorphosis, a specific theory of evolutionary progress in the human phylogenetic line. An essential role here is the *biosocial selection* defined as a differential survival and reproductive success related to diverse degrees of adaptation to social and labor contacts. Biosocial selection is considered by Yefimov to be an important driving force in the transient period of evolution at the demarcation line between ape and man. Yefimov is convinced that biosocial selection has individual and group aspects, each resulting in differential success of hominid bands. Biosocial selection thus may be the central force of epimorphosis, responsible for the transcendence of the limits of conventional aromorphosis. Yefimov shares the conviction with many Soviet biologists and anthropologists that natural selection ceased to be an effective factor of evolutionary change with the appearance of true *Homo sapiens*. This standpoint was firmly established by Kremiansky's (1941) paper on "self elimination" of natural selection in the evolution of man. This self elimination, due to the leading role of labor, has been accepted by numerous Soviet biologists -- for example, a neo-Darwinian, Bystrov (1957) elevated this principle to the central idea of human evolution -- who otherwise represented different views on evolution and its causes.

Yefimov treats the idea of epimorphosis as a major theoretical achievement, crucial to the understanding of the ascent of Man. Moreover, Schmalhausen's concepts blended with dialectical and historical materialism to form a unified view of evolutionary biology and Marxism.

Acknowledgments

I am grateful to the Field Museum of Natural History for the 1987 tenure of the Visiting Scientist position during which most of this paper was written. It is a pleasure to acknowledge my debt to Drs. Matthew H. Nitecki, David M. Raup and David Joravsky, for the insightful comments on my paper. I also thank Miss Elizabeth Moore for her patient typing of the manuscript.

References

Adams, M. B. 1980. Severtsov and Schmalhausen. Russian morphology and the evolutionary synthesis. In *The evolutionary synthesis*, ed. E. Mayr and W. B. Provine, 193-225. Cambridge: Harvard University Press.

Ayala, F. J. 1974. The concept of biological progress. In *Studies in the philosophy of biology. Reduction and related problems*, ed. F. J. Ayala and T. Dobzhansky, 339-55. Berkeley and Los Angeles: University of California Press.

Bystrov, A. P. 1957. *Proshloe, nastoyashchee, budushchee cheloveka* (The past, present and future of man). Leningrad: Medgiz.

Darwin, C. 1859. *On the origin of species by means of natural selection or the preservation of favored races in the struggle for life*. London: J. Murray.

Davitashvili, L. Sh. 1956. *Ocherki po istorii ob evoliutsionnom progriessie* (Essays on the history of evolutionary progress). Moskva: Izdat. Akademii Nauk.

Davitashvili, L. Sh. 1961. Uchenie ob evoliutsionnom progriessie i zadachi sovremiennoy biologii (Theory of evolutionary progress and the tasks of contemporary biology). *Trudy Inst. paleobiol.* 4:3-26.

Davitashvili, L. Sh. 1972. *Uchenie ob evolutsionnom progresse. Teoriya aromorfoza* (Problem of the evolutionary progress. The theory of aromorphosis). Tbilisi: Metzniereba Publishers.

Davitashvili, L. Sh. 1978. *Evoliutsionnoe uchenie* (The theory of evolution). 2 vols. Tbilisi: Metzniereba Publishers.

Franz, V. 1935. Zum jetzigen Stand der Theorie vom biotechnischen Fortschritt in der Pflanzen und Tiergeschichte. *Biologia Generalis* 19:3.

Granin, D. 1987. *Zubr. Poviest* (Bison. A novel). *Novy Mir* 1:19-95; 2:7-92.

Huxley, J. S. 1942. *Evolution. The modern synthesis*. London: Allen and Unwin.

Kapralova, T. I., and T. A. Lukina. 1970. K voprosu o biotiechnicheskom progriessie (Contribution to the biotechnical progress). In *Tieorieticheskie voprosy progriessivnogo rozvitia zhivoi prirody i tiekhniki* (Theoretical problems of progressive development in living nature and technology), ed. K. M. Zavadsky, and Yu. S. Meleshchenko, 65-74. Leningrad: Nauka.

Kremiansky, V. I. 1941. Perekhod ot vieduschei roli yestie striennogo otbora k vedushchey roli truda (Transition from the leading role of natural selection into the leading role of the labor). *Uspekhi Sovr. Biologii* 14, 2:356-71.

Lamarck, J.-B. 1809. *Philosophie zoologique, au exposition des considerationes relatives a l'histoire naturelle des animaux*. Paris: Dentu.

Matveev, B. S. 1967. Obzor novykh dannykh o proiskhozdenii i pytyakh evolutsii pozvonochnykh zhivotnykh (A review of the new data on the origin and evolutionary trends in the vertebrates). In *Glavnyje napravlienija evolutsionnogo processa. Morfobiologicheskaya tieorya evoliutsii* (Main trends of evolutionary process. Morphobiological theory of evolution), A. N. Severtsov, 184-201. Moskva: Izdat. Moskovskogo Universitieta.

Miklin, A. M. 1970. O kritieriakh progriessa v zhivoi prirode i tiekhnikie (On the criteria of progress in living nature and in technology). In *Tieorieticheskie voprosy progriessivnogo rozvitia zhivoi prirody i tiekhniki* (Theoretical problems of progressive development in living nature and technology), ed. K. M. Zavadsky and Yu. S. Meleshchenko, 101-26. Leningrad: Nauka.

Miklin, A. M. 1983. Krierii progriessivnoy evoliutsii (Criteria of progressive evolution). In *Razvitie evoliutsionnoi teori v SSSR* (The development of theory of evolution in the USSR), ed. S. R. Mikulinskii and Yu. I. Polyansky, 358-64. Leningrad: Nauka.

Mirzoyan, E. N. 1983. Uchenye A. N. Severtsova o glavnikh napravlenijakh evoliutsionnogo processa (The doctrine of A. N. Severtsov on the main trends of evolutionary process). In *Razvitie evoliutsionnoi teorii v SSSR* (The development of theory of evolution in the USSR), ed. S. R. Mikulinskii and Yu. I. Polyansky, 348-58. Leningrad: Nauka.

Nägeli, C. 1884. *Mechanisch-physiologische Theorie der Abstammungslehre*. Leipzig: Oldenbourg.

Paramonov, A. A. 1967. Puti i zakonomiernosti evoliutsionnogo prociessa (Routes and modes of evolutionary process). In *Sovremiennyie problemy evoliutsionnoi tieorii* (Contemporary problems of evolutionary theory), ed. Z. I. Berman and K. M. Zavadsky, 342-441. Leningrad: Nauka.

Schmalhausen, I. I. [1939] 1983. *Puti i zakonomernosti evolutsionnogo prociessa* (Routes and modes of the evolutionary process). 2d ed. Moskva-Leningrad: Nauka.

Severtsov (Severtzoff), A. N. 1931. *Morphologische Gesetzmässigkeiten der Evolution*. Jena: Gustav Fischer.

Severtsov, A. N. 1939. *Morfologicheskie zakonomiernosti evoliutsi* (Morphological Laws of Evolution). Moskva-Leningrad: Izdat. Akademii Nauk.

Tachtajan, A. L. 1951. *Puti prisposobitielnoi evoliutsii rastienii* (Routes of adaptive evolution of plants). *Botan. Zhurn.* 36, 3:231-39.

Tachtajan, A. L. 1954. *Voprosy evoliutsionnoi morfologii rastienii* (Problems of Evolutionary Morphology of Plants). Leningrad: Nauka.

Tatarinov, L. P. 1976. *Morfologicheskaya evoliutsia teriodontov i obschie voprosy filogenietiki* (Morphological evolution of theriodonts and general problems of phylogenetics). Moskva: Nauka.

Yablokov, A. V. 1968. O raznykh formakh progriessivnogo razvitya v organicheskoi prirode (On the different forms of progressive evolution in the organic nature). In *Problemy evoliutsii*, I (Problems of evolution, I), ed. N. N. Vorontsov, 89-115. Novosibirsk: Nauka.

Yefimov, Yu. I. 1972. Epimorfoz i nieogranichennyi progriess (Epimorphosis and unlimited progress). In *Zakonomiernosti progriessivnoi evoliutsii* (Laws of progressive evolution), ed. K. M. Zavadsky, 105-18. Leningrad: Izdat. Akademii Nauk.

Yefimov, Yu. I. 1981. Filosofskiye problemy tieorii antroposociogienieza (Philosophical problems of the anthroposociogenesis thesis). Leningrad: Nauka.

Yefimov, Yu. I. 1983. Razrabotka teorii anthropogienieza (The development of the anthropogenesis theory). In *Razvitie evoliutsionnoi teorii v SSSR* (The development of theory of evolution in the USSR), ed. S. R. Mikulinskii and Yu. I Polyansky, 435-48. Leningrad: Nauka.

Zavadsky, K. M. 1953. Ob uchenii akadiemika A. N. Severtsova (On the theory of Academician A. N. Severtsov). *Vestnik Leningradskogo Gos. Universitieta* 7:3-23.

Zavadsky, K. M. 1958. K ponimaniyu progriessa v organicheskoi prirode (Contribution to the understanding of progress in organic nature). In *Problemy razvitiya v prirode i obschestvie* (Problems of development in nature and society), ed. B. A. Chagin, 79-120. Leningrad: Izdat. Akademii Nauk.

Zavadsky, K. M. 1967. Probliema progriessa zhivoi prirody (Problem of the progress in the living nature). *Voprosy filosofii* 9:124-36.

Zavadsky, K. M. 1970. K problemie progriessa zhivykh i tekhnicheskikh sistiem (On the progress in living and technical systems). In *Tieorieticheskie voprosy progriessivnogo rozvitiya zhivoi prirody i tiekhniki* (Theoretical problems of progressive development in living nature and technology), ed. K. M. Zavadsky and Yu. S. Meleschenko, 3-28. Leningrad: Nauka.

Zavadsky, K. M. 1971. K issledovaniju dvizhushchikh sil arogienieza (On the study of dynamic forces of arogenesis). *Zhurn. Obsch. Biologii* 52:515-29.

Zavadsky, K. M. 1972. O prichinakh evoliutsii v storonu arogienieza (On the causes of evolution towards arogenesis). In *Zakonomiernosti progriessivnoi evoliutsii* (Laws of progressive evolution), ed. K. M. Zavadsky, 135-48. Leningrad: Izdat. Akademii Nauk.

Zavadsky, K. M., and E. I. Kolchinsky. 1972. Evoliutsionno-biologicheskaya formacija. Predvaritelnaye opredielienie poniatii (Evolutionary-biological formation. A preliminary definition). In *Zakonomiernosti progriessivnoi evoliutsii* (Laws of progressive evolution), ed. K. M. Zavadsky, 149-56. Leningrad: Izdat. Akademii Nauk.

Zavadsky, K. M., and E. I. Kolchinsky. 1977. *Evoliutsia evoliutsii* (Evolution of evolution). Leningrad: Nauka.

EMPIRICAL APPROACHES

Evolutionary Progress and Levels of Selection

John Maynard Smith

The concept of progress has a bad name in evolutionary biology. There are a number of good reasons for this. In the first part of this essay, I review these reasons: in the second part, I suggest a way of classifying the stages through which life has passed. Although biologists may be reluctant to see these stages as representing "advance," they are progressive in one sense: the sequence in which they have occurred is not arbitrary, since each stage was a necessary precondition for the next. But first, what are the reasons for distrusting the idea of progress?

The first is semantic. How shall we define progress? There is an obvious danger of anthropomorphism: the more similar an organism is to us, the more advanced it seems. Historically, the idea of evolutionary progress derives from the medieval ladder of life, although to be fair, it was not assumed that we stood at the top of the ladder. There were angels and archangels above us, as there were beasts below. The snag, of course, is that life is better represented by a branching tree than by a linear progression. Does the claim of progress mean that there has been progress in all the branches, or, more plausibly, in at least one?

There is a natural wish to measure progress in terms of morphological complexity: there must be some sense in which an oak tree or an elephant is structurally more complex than anything that existed, say, 10^9 years ago. However, even attempts to measure morphological difference, without any implication of progress, have proved to be full of pitfalls. When comparing different taxa, it seems that the best one can do is measure the number of cell types present. An easier measure of complexity is the amount of coding DNA present in the genome. Some estimates are given in Table 1. There is some slight tendency for organisms that we think of as structurally more

Table 1

DNA content per haploid genome

	Genome size (pg)	Percent of genome coding for protein	Coding DNA (pg)
Escherichia coli	0.004	100.0	0.4
Saccharomyces cerevisiae	0.009	69.0	0.62
Caenorhabditis elegans	0.088	25.0	2.2
Drosophila melanogaster	0.18	33.0	5.9
Homo sapiens	3.5	20.0	70.0
Triturus cristatus	19.0	3.0	57.0
Protopterus aethiopicus	142.0	0.8	114.0
Arabidopsis thaliana	0.2	31.0	6.2
Fritillaria assyriaca	127.0	0.02	2.5

Data from Cavalier-Smith 1985. 1 pg of DNA corresponds to approximately 10^9 base pairs. The values of the percent DNA coding for protein are the averages of different estimates, and are only approximate.

complex to have more coding DNA, but it is not very striking. In any case, it is well to remember that the genome is best thought of as a set of instructions for making an organism, and not as a description of the adult structure. One form of progress, therefore, could be the evolution of a more efficient (in terms of quantity of information) set of instructions for making the same structure: if so, progress would be accompanied by a decrease in coding DNA.

A second difficulty is that our theory of evolution does not predict an increase in anything. At first sight, Fisher's (1930) "fundamental theorem of natural selection" might seem to predict an increase in "mean fitness," but it would be a mistake to think that there is any quantity that necessarily increases, as entropy increases in a closed physical system. Thus consider the following imaginary experiments. First, a physicist is provided with any information he wants about the state of a closed physical system at two points in time, but is not told which state is the earlier one. Provided that entropy was not at a maximum on both occasions (that is, provided that something interesting had happened), she could say which state is the earlier one. Now consider a closed biological system: by "closed" I mean that no living organisms enter or leave the system. There is nothing in Fisher's theorem that would enable a biologist to say which state was earlier. Thus even for a single species in a constant physical environment, there can be a continuous cycle of gene frequencies if fitnesses are frequency dependent. Hence, given data on a population at two moments, one could not tell which was earlier. For evolution in multi-species communities, cycles and other more complex dynamics are likely to be the rule. For evolutionary changes, therefore, it is hard to put an arrow on time, let alone demonstrate the inevitability of progress.

Finally, there is little empirical support for the concept of progress. Others will discuss the fossil record from this point of view, so I will consider the matter from the opposite extreme -- the study of evolution in vitro (Orgel 1979). If RNA molecules are replicated in a test tube by the viral enzyme, Qß replicase, it is possible to follow the base sequence and secondary structure of the population of molecules. More or less independently of the starting point (that is, of the base sequence of the RNA molecule used as an initial primer of the system), the end point is a rather small molecule, some 200 bases long, with a particular sequence and structure that enable it to be replicated particularly rapidly. In this simple and well-defined system, natural selection does not lead to continuing change, still less to anything that could be recognized as an increase in complexity: it leads to a stable and rather simple end point. This raises the following simple, and I think unanswered, question: What features must be present in a system if it is to lead to indefinitely continuing evolutionary change?

To summarize, progress is hard to define. If we equate progress to increase in complexity, the best we can do is to measure complexity in terms of the quantity of coding DNA, or, in eukaryotes, by the number of distinct cell types. Theoretically, there is no quantity, analogous to entropy, that necessarily increases in an evolving population. Observationally, natural selection in the simplest systems does not lead to indefinitely continuing change, let alone to progress. It is true that, on the largest scale, evolution does seem to have given rise to increasingly complex organisms. But, since the first living things were necessarily simple, it is not surprising that the most complex things alive today are more complex than their first progenitors.

This is a rather boring conclusion. It is more fruitful to attempt to categorize the levels of complexity through which life has passed. In doing this, I use Darwinian theory to classify levels of complexity, but I do not

Table 2

Levels of complexity, differing in the organization of the genetic material

1. Replicating molecules	5. Multicellular organisms
2. Populations of molecules in compartments	6. Demes; social groups
	7. Species
3. Prokaryotic cells	8. Groups with cultural inheritance
4. Eukaryotic cells	

claim that the various levels are predictable from that theory. Darwin's theory can be summarized as follows: Any population of entities with the properties of multiplication (one entity can give rise to many), variation (entities are not all alike, and some kinds are more likely to survive and multiply than others), and heredity (like begets like) will evolve. Evolution on earth has occurred because living organisms have these properties. A major problem in current evolutionary theory is to identify the relevant entities. For Darwin, they were individual organisms. Today, they have been variously identified as genes, genomes, organisms, demes, species, communities and even, somewhat illogically, as the whole biosphere. In what follows, I shall refer to entities with the three necessary properties as "units of evolution." I prefer this phrase to "units of selection," because selection often acts on objects that lack heredity, and so can not evolve by natural selection. Units of evolution can be expected to evolve characteristics that ensure their own survival and reproduction.

During evolutionary time, the entities that constitute units of evolution have passed through a number of levels of complexity. A somewhat speculative series of stages is shown in Table 2. The table is speculative in two ways. There is no direct evidence of the existence of stages 1 and 2, but we infer that they existed because it is hard to see how life could get started otherwise. There is no doubt that stages 3 to 8 exist, but it is arguable whether the last three should be regarded as units of evolution.

A classification of this kind is, of course, a truism. It becomes interesting if we ask the following questions: (i) What is the nature of the genetic information that is passed from generation to generation at each stage? (ii) How is the integrity of that information protected against selection at lower levels? (iii) How did natural selection bring about the transition from one stage to another, since, at each transition, selection for "selfishness" between entities at the lower level would tend to counteract the change.

A full answer to these questions would constitute a solution to most of the outstanding problems in evolutionary biology. I can make only a few brief suggestions. My main aim is to bring out the analogies between the different transitions. In each case, the difficulty is the same: how is it that selection at a lower level does not disrupt integration at a higher one?

Replicating Molecules

It is widely believed that the first Darwinian entities were RNA-like molecules, replicating by complementary base pairing. Orgel has recently suggested that RNA itself may not be primitive, and that in the first replicators ribose was replaced by some simpler compound: what is primitive is replication that is dependent on base pairing. In the absence of enzymes, replication accuracy would be low, and this would limit the length of the molecules whose sequence could be maintained by selection to, very approximately, 10-100 bases. This would rule out the evolution of a genetic message long enough to code for the enzymes needed to improve the accuracy of replication. This difficulty led Eigen and Schuster (1979) to suggest the evolution of "hypercycles." Essentially, a hypercycle is a collection of molecules of different kinds, functionally arranged in a cycle, ... A-B-C-D-A-B ..., so that the rate of synthesis of each molecule in the cycle is accelerated by the concentration of the molecule immediately preceding it. Such a system can be chemically stable, and can transmit a greater total quantity of information than a single molecule. However, selection will not readily incorporate mutations that increase the efficiency of the cycle as a whole.

Thus suppose that a mutation in molecule A makes it better at stimulating the synthesis of B. This would improve the efficiency of the cycle as a whole, but the mutant would not increase under natural selection. By analogy, a mutation in grass that makes it better for nourishing antelopes would not spread. Selection at the level of individual molecules need not lead to improvement of the higher level system, the hypercycle. This difficulty can be overcome by enclosing the population of molecules within a "compartment," leading to the next stage.

Populations of Molecules in Compartments

If populations of replicating molecules are enclosed -- in spherical membranes, or in the droplets of a coacervate as suggested by Oparin (1957), or in proteinoid microspheres as suggested by Fox (1984) -- the difficulty is partly overcome. A mutant A molecule that accelerates the synthesis of B will, because of the cyclic arrangement, accelerate its own synthesis, whereas in the absence of enclosure it would accelerate the synthesis of other nonmutant A molecules. However, if the compartments, rather than the individual molecules, are to be the units of evolution, they must possess the properties of multiplication, variation and heredity. They will have heredity, because when a compartment splits, the molecular species it contains will be passed on. But it must also be true that a compartment must grow and divide more rapidly if the molecules it contains replicate faster.

Even with between-compartment selection, however, it is not clear how far the hypercycle is proof against selfish mutations at the molecular level. Szathmary and Demeter (1987) have recently argued that further evolution at the compartment level requires an additional feature, which they refer to as the "stochastic corrector." They require that the numbers of molecules of each kind in a compartment be small. This has the effect that chance differences between compartments arise at division, thus making selection at the higher level more effective. This is a feature of all transitions between levels: we shall meet it again when discussing the origin of multicellular organisms, and the effectiveness of interdemic selection.

Selection between compartments is more effective if the number of molecules is small. However, if molecules are partitioned randomly between daughters, there is a risk of a daughter receiving no molecules of a particular type. This risk can be avoided by joining all the types, which we can now call genes, end to end on a chromosome, of which exactly one copy is passed to each daughter cell. This is possible only when the accuracy of replication is high enough. Initially, when replication is inaccurate, it must be possible for selection, within compartments, to act separately on each type, to prevent the accumulation of errors, and this requires a population within the cell. We are still some way from understanding these transitions, from replicating molecules to populations of interacting molecules within compartments, and thence to cells with chromosomes.

Prokaryotic Cells

During most of the history of life, the predominant units of evolution have been prokaryotic cells whose genetic information is carried in a single circular chromosome. The theory of prokaryotic evolution is in its infancy. We have to learn the evolutionary significance of bacterial conjugation and transformation, and the roles of phages, plasmids and transposons. In the context of the evolution of increasing complexity, Cavalier-Smith (1981) has argued that the crucial limitation lies in the fact that there can be only one point of initiation of replication per chromosome, if one copy is to be transmitted to each daughter cell. This places an upper limit on the quantity of genetic information per cell.

Eukaryotic Cells with Sex

The symbiotic origin of eukaryotic cells brought together genetic information from several different prokaryotic lineages. A more important factor in permitting an increase in complexity is the existence of several chromosomes, each with multiple sites of initiation of replication, made possible by the evolution of mitosis. Since natural selection lacks foresight, we cannot argue that mitosis evolved because it made possible an increase in complexity. However, a change which speeded up the replication of a given quantity of DNA would also permit a future increase in that quantity.

It is the origin of meiosis and syngamy, however, that raises the real difficulties for Darwinian theory, because the genetic information present in an individual is not passed as a unit to its offspring. This has led Williams (1966) and Dawkins (1976) to argue that, since the gene is the largest unit that is transmitted without recombination for reasonably long periods, we must see the genes as the unit of selection. I think this is correct (although, for the reasons given above, I would prefer the term "unit of evolution"). For many purposes, however, we can safely think in terms of individuals. The reason is that, almost always, there is no within-individual selection between genes at a locus: in other words, meiotic drive is very rare. Consequently, a gene will increase in frequency only if, on average, it increases the fitness of its carriers. "On average" implies averaging over environments, and over the genes present in the population at other loci. Therefore, although genomes are not replicated as units, we can understand why natural selection leads to adaptation at the individual level.

To achieve individual adaptation, it is not sufficient that selection should, typically, act on individuals: it is also necessary that selection between genes within individuals should be unimportant. The argument, by Mayr (1963) and others, that because the genes in an organism are coadapted there must be something inadequate in "beanbag genetics," is a misunderstanding. Of course the genes in an organism are coadapted, in the sense that a set of rabbit genes or of mouse genes work well together, but a set consisting half of rabbit genes and half of mouse genes would not. This is so because a gene only spreads under selection if its effects are coadapted to the other genes in the population. Coadaptation is a consequence of beanbag genetics. In most species, there is no genetic entity much larger than a gene which is transmitted as a unit for long enough to serve as a unit of evolution.

All this seems tolerably clear. The real difficulty arises when we ask why meiosis and syngamy arose in the first place, and why they are maintained in most eukaryotes. I will return to this problem when discussing the species level.

Multicellular Eukaryotes

Higher animals and plants differ from protists in that there are, in the adult, many copies of the genetic information, one (or two in diploids) in each cell. Some implications of this fact are discussed in a forthcoming book, *The*

Evolution of Individuality, by Leo Buss. For example, he thinks that the early
segregation of the germ line evolved because it limits between-cell, within-
individual competition for the privilege of becoming a gamete. However, I
suspect that in this context the most significant thing about multicellular
organisms is that, typically, they reproduce through single cells. As in the
case of the stochastic corrector model discussed above, this increases the
effectiveness of selection at the higher (organism) level relative to the lower
(cell). This raises an interesting question concerning clonal organisms that
reproduce by multicellular propagules such as bulbils or rhizomes. I can see
two possibilities. One is that such organisms are doomed to death through the
accumulation of selfish mutations if they do not periodically revert to
reproduction through seeds or eggs. The other is that the cells of the
propagule are descended from a single cell in the relatively recent past.

Demes and Social Groups

If individuals are populations of cells, and yet can evolve individual-level
adaptations, can not demes, which are populations of individuals, evolve group-
level adaptations? In particular, can they not evolve behavior that is
favorable to group survival? This was the form of group selection proposed by
Wynne-Edwards (1962). If new groups are established by single fertilized
females, then genes that are advantageous to the group, but are selected
against within groups, can sometimes spread. However, it is rare for higher
organisms to have an appropriate population structure. It has proved to be
more fruitful to analyze social behavior either from a gene-centered or an
individual-centered viewpoint, allowing in either case for the genetic
relatedness of interacting individuals (Hamilton 1964). In the gene-centered
approach, we ask whether the phenotypic effects of a gene are such as to
increase the number of copies of that gene in future generations. In the
individual-centered approach, we ascribe to each individual an "inclusive
fitness."

There are, however, occasions on which it seems reasonable to interpret
the behavior of individuals solely in terms of the contribution they make to
the success of the group, rather as one might discuss the contribution of liver

cells to individual survival. For example, Oster and Wilson (1978) discussed caste differentiation in ants in this way. The justification, I think, is that some social insects have evolved so far along the road to sociality that some individuals can ensure the transmission of their genes only by contributing to colony success. A worker ant or bee would gain nothing by leaving the colony. However, this is always a risky approach: as Trivers and Hare (1976) pointed out, workers in social insect colonies can increase their inclusive fitness by distorting the colony sex ratio, or even by producing sons parthenogenetically. It seems likely that, in all social animals, the opportunities for within-group, between-individual selection are substantial.

Species

The gene pool can be thought of as carrying the genetic information of a species, rather as the genome does of an individual. When a species splits in two, the daughter species carry most of the ancestral genes, so that "species heredity" is ensured. Species, therefore, are candidates as units of evolution. The snag, of course, is that within-species, between-individual selection is ubiquitous, and strong enough to prevent species-level adaptations.

Sexual reproduction may constitute an exception to this rule. Theory suggests that sex can confer an advantage on the population by accelerating evolution and by delaying the accumulation of deleterious mutations. The taxonomic data suggest that parthenogenetic varieties are short-lived in evolutionary time. But what of the strength of individual selection, particularly in view of the twofold cost of meiosis? One answer is that many sexual organisms have been sexual for so long, and have so many secondary adaptations associated with sex that they cannot readily abandon it. If there are no mutations to sexuality, then selection is powerless. However, there are many taxa in which closely related sexual and parthenogenetic taxa live side by side. Williams (1975) argued that the existence of such cases shows that there must be some short-term advantages to sex. It is on these cases that research should concentrate.

Even if we accept that species-level selection is important in maintaining sex, there are two reservations to be made. The first is that meiosis and syngamy could not have evolved in the first place by such a mechanism. The second is that various processes subsidiary to sex still require a short term explanation. In particular, the rate of recombination requires such an explanation, because we know that there is heritable variation within species for the frequency and the location of chiasmata.

Groups with Cultural Inheritance

Human society is the final level of complexity so far achieved by living organisms. It depends on language, which has the consequence that information is transmitted, horizontally and between generations, by cultural rather than genetic means. The formal differences between cultural history and genetic evolution are profound, but it is intriguing that history reveals the same conflict between individual and group interests as has been fundamental in evolution.

Conclusion

In this brief account, I have omitted some levels of organization -- those of the ecological community and of the biosphere as a whole -- which are customarily included in hierarchical accounts of biological organization. This is because I do not see any sense, however stretched, in which these levels can be treated as Darwinian units of evolution. They do, however, reveal properties of stability and persistence that we find hard to explain. No Darwinist could accept the "Gaia" hypothesis, according to which the earth is analogous to a living organism, because the earth is not an entity with multiplication, variation and heredity. However, we should not be too contemptuous of that idea, logically flawed as it is, until we can give a better account of the long-term stability of the biosphere than is at present possible.

I do not think that progress is an inevitable consequence of evolution by natural selection. However, one can recognize in the evolution of life several revolutions in the way in which genetic information is organized. In each of these revolutions, there has been a conflict between selection at several levels.

The achievement of individuality at the higher level has required that the disruptive effects of selection at the lower level be suppressed. It is this common feature of these transitions, which are in other ways as disparate as that from molecule to cell and from protist to metazoan, which makes it worthwhile to compare them.

References

Buss, L. *The evolution of individuality*. Forthcoming.

Cavalier-Smith, T. 1981. The origin and early evolution of the eukaryote cell. In *Molecular and cellular aspects of microbial evolution*, ed. M. J. Carlisle, J. F. Collins and B. E. B. Moseley. Society for General Microbiology Symposium 32:33-84. Cambridge: Cambridge University Press.

Cavalier-Smith, T. 1985. *The evolution of genome size*. Chichester: Wiley.

Dawkins, R. 1976. *The selfish gene*. Oxford: Oxford University Press.

Eigen, M., and P. Schuster. 1979. *The hypercycle*. Berlin: Springer-Verlag.

Fisher, R. A. 1930. *The genetical theory of natural selection*. Oxford: Clarendon Press.

Fox, S. W. 1984. Protenoid experiments and evolutionary theory. In *Beyond neo-Darwinism*, ed. M.-W. Ho and P. T. Saunders. New York: Academic Press.

Hamilton, W. D. 1964. The genetical evolution of social behavior. *Journal of Theoretical Biology* 7:1-52.

Mayr, E. 1963. *Animal species and evolution*. Cambridge: Harvard University Press.

Oparin, A. I. 1957. *The origin of life*. New York: Academic Press.

Orgel, L. E. 1979. Selection in vitro. *Proceedings of the Royal Society, London*. B205:435-42.

Oster, G. F., and E. O. Wilson. 1978. *Caste and ecology in the social insects*. Princeton: Princeton University Press.

Szathmary, E., and L. Demeter. 1987. Group selection of early replicators and the origin of life. *Journal of Theoretical Biology 128:463-86*.

Trivers, R. L., and H. Hare. 1976. Haplodiploidy and the evolution of the social insects. *Science* 191:249-63.

Williams, G. C. 1966. *Adaptation and natural selection*. Princeton: Princeton University Press.

Williams, G. C. 1975. *Sex and evolution*. Princeton: Princeton University Press.

Wynne-Edwards, V. C. 1962. *Animal dispersion in relation to social behaviour*. Edinburgh: Oliver and Boyd.

Two Constraints on
the Evolution of Complex Adaptations
and the Means for their Avoidance

William C. Wimsatt and Jeffrey C. Schank

The concept of progress in evolution has had a long and checkered history, all too often because it has been invoked to glorify or to justify certain aspects of the human condition, and more recently, because it has seemed to have little to do with, or even to conflict with the accepted mechanisms of evolutionary change. Nineteenth century writers took it for granted that evolution was progressive, that progress implied a change for the better, and that man, particularly Western European man and culture, represented the apex of this progressive change. "Progress" became a tool for justifying the free market, for colonial domination and exploitation of non-Western "primitive" societies, and for the manipulation and exploitation of our natural and biological environment.

Twentieth century evolutionists have, for the most part, properly rejected all of these ideas. Along with this however has gone a suspicion of any claims for systematic trends (however nonevaluative and nonprogressive in the nineteenth century sense) in the course of evolution. George Williams (1966), in a widely influential work which provided the watershed of recent reductionistic "genic selectionist" interpretations of evolutionary theory, argued that on a variety of plausible criteria, evolutionary products showed no unambiguously demonstrable signs of secular, much less progressive change and that notions of long range progress had no proper role to play in evolutionary theory.[1] Higher levels of adaptation and evolutionary conceptions of progress seemed ephemeral when treated with the reductive skepticism which Williams made popular--a stream which became a river flooding the varied territories of evolutionary biology.

This high tide of reductionism has by now abated somewhat. Selection at the level of gene complexes (Lewontin 1974; Wimsatt 1980, 1981), of conspecific groups, (Wilson 1975, 1980; Wade 1976, 1977, 1978), or even of the species (Gould 1982; Arnold and Fristrup 1982) or the ecosystem (Wilson and Sober 1987) has seen new more convincing empirical demonstrations, theoretical models, and conceptual formulations. In at least the second and third areas, these ideas have won a growing cadre of new adherents (Wimsatt 1980, 1981, and papers in Brandon and Burian 1983), and the first claim has never been without its defenders. The recent growth of interest in hierarchical models of selection and of evolution has produced a context in which "progress" may seem like somewhat less of a naughty word, at least if spoken softly and understood as carrying none of its nineteenth century evaluative implications. Thus, for example, if there are multiple levels at which selection acts, and higher levels evolved later than lower ones, a recognition of a kind of progressive change (whether for better or worse!) seems inevitable. In fact, one of us (Wimsatt) would like to argue this case on another occasion (there are some interesting twists which strengthen the argument that higher levels of selection will emerge), but this is not our present concern.

We wish to argue for another kind of progress in evolution. Williams's critical discussion in 1966 did not exhaust all of the possibilities for progress as *a systematic increase (or decrease) in some biologically significant property*, but any such discussion requires some prior qualifications. These qualifications were probably intended by the authors Williams criticized, but a failure to make them renders any proposal for a criterion of evolutionary progress an easy target for attack. Given the diversity of kinds of evolutionary change in different lineages, almost anything can happen and will. For example, a claim for an evolutionary increase in complexity will find ready counterexamples in trends towards increasing simplicity in some lineages (Williams 1966), and this same kind of diversity in evolutionary trends is bound to bedevil any criterion of progress. Thus the following qualifications seem necessary:

By systematic change, we mean tendencies. Therefore, the change in question would generally (1) not be monotonic in any lineage, or (2) universal across lineages; nor (3) is it even clear that the overall average value of the property across all lineages must be increasing. This greatly weakened notion of progress could still be interesting if (4) the property in question had sufficient biological interest, (5) its maximum value increased over time, and (6) there were a tendency for its value to increase in many lineages. There is such a property, and we believe that it meets all of these conditions.

We wish to argue for an increase in the possibilities for, and capability of, producing and maintaining a larger variety of more complex adaptations through evolutionary time, though what we will actually demonstrate, if successful, is somewhat more modest. This capability cannot be observed directly, though increases in the maximum observed morphological, behavioral, developmental, and life-cycle complexity over time would all presumably be consequences of its growth. They are not good evidence for it, however, for several reasons: (1) There usually is a variety of possible alternative explanations for these changes, many more local, special and ecological in character, and there is a strong preference for explanations which (like these alternatives) are more closely tied to microevolutionary mechanisms. (2) For this reason, there would be fundamental disagreement over the level at which an explanation should be offered, and more strongly, whether any single explanation is called for. (3) Because we do not yet have any widely agreed upon metrics for measuring the complexity of an organism in any of these categories, it is not clear how such claims are to be evaluated. (4) Since there seem to be myriad exceptions to such claims, even if they are not understood universally, it is not clear how many exceptions are required to falsify the claim, or which exceptions should or should not count. (5) Finally, we have neither the data nor the standards (in particular, as to what should count as the appropriate null hypothesis) for evaluating them. Thus, suppose that evolution involves random changes in the parameter in question in different lineages, and there is an increase in the number of lineages extant at any given time. In that case we would expect to see the maximum value of the parameter increase over time, and with a sufficient number of lineages under consideration, one would be bound to find a systematic increase in that property in some fraction of them.

Given this thicket of problems, why should one try to argue for such a conclusion? There is at least one remaining basis for argument, and this basis provides more structure for the discussion so that it need not be hopelessly vague or inconclusive. We believe that it can be shown on theoretical grounds that such progress results as a consequence of general results of population genetics and certain microevolutionary processes and constraints on the evolution of the causal structure of developmental interactions. In particular we argue that *selection acting over evolutionary significant periods of time will, in a variety of circumstances and in a variety of ways, act to change the architecture of gene expression and other properties of the genome so as to decrease the average amount of genetic load per locus, thus allowing the maintenance, at or near fixation, of alleles at a larger number of loci.*[2] This does not entail, but should allow and make more probable the evolutionary development and maintenance of a larger number of more complex adaptations, where complexity is measured by the number of genes involved specifically in the production of that or those adaptations.[3] The first claim, the one which we will defend, is related to microevolutionary mechanisms, but operates on a macroevolutionary time scale. It thus cannot be tested (except for the plausibility of its mechanisms) over microevolutionary time. But if the arguments to be presented here are correct, an increase in the complexity of adaptation in at least some species lines over longer evolutionary periods of time should be virtually inevitable.

Claims for progress of this sort should be something more than a hunting license to search for cases which fit the rule, an abuse which would constitute another kind of misapplication of the "adaptationist program" (as criticized by Gould and Lewontin 1979). Given the richness and variety of kinds of organisms and interactions, this procedure should make any responsible biologist suspicious that this is but an exercise in opportunistic curve-fitting which cannot fail, and which therefore, at another level, cannot succeed. But given the analysis below, it should be possible to say what kinds of conditions facilitate this growth in complexity, and the mechanisms proposed should be amenable to study through computer simulation, a task which we have started and plan to continue. We hope that the theoretical justification for these claims and the structure which the analysis provides will make them more testable and better grounded than prior attempts.

The Characteristics and Desiderata for Evolutionary Constraints

Notions of evolutionary progress or change are productively discussed in terms of the closely related notion of an evolutionary constraint, though not every evolutionary constraint would promote evolutionary change, and some would seem to be inconsistent with it. It is common to think of constraints in a negative fashion -- as preventing things from happening, and thereby reducing the variety found in nature. But if the process of producing variation is open-ended, the introduction of constraints can channel the variation, and by directing it, produce much further or deeper exploration in a given direction than would otherwise be possible. Constraints can thus play a creative and, in one sense, ultimately progressive role. This is a deep truth, not only about evolution, but about problem-solving and exploration in general. It is why Darwin was right in 1859 when he saw natural selection as a creative force, and why his critics who saw selection only as playing a negative role by eliminating variety were wrong.

What is an evolutionary constraint? To be counted as a constraint, the condition must have some generality, and not apply merely to the extinction, stasis, or direction of change of a single lineage or small set of lineages meeting very special conditions. With this qualification, various things have been called constraints, including both absolute requirements (Darwin's principles, below) or requirements which can be met in various degrees, and in which the degree in question can determine or constrain the rate of an evolutionary process. Thus, the degree of linkage of an allele with alleles at other loci with which it has epistatic interactions can determine the maximum rate at which it can proceed to fixation, or indeed, whether it will ever do so (Lewontin 1974). The principle of quasi-independence (below) also allows such a degree interpretation. Finally, there are constraints which specify things that cannot happen, or will not happen, because they are sufficiently improbable. The neo-Darwinian bias against evolution through the frequent occurrence of adaptive macromutations has this kind of motivation for its most convincing justification. Those who have recently argued for a role for macromutations in evolution (Arthur 1984; Wimsatt 1986; Schank and Wimsatt in press) take it for granted that the frequency of adaptive mutations declines rapidly with their size and complexity, and that adaptive macromutations would be expected to be exceedingly rare, even on an evolutionary time scale.[4]

It is also desirable for several reasons that the proposed evolutionary constraint have a basis in microevolutionary processes: (1) Such a constraint

becomes much more readily testable than one which is only realized on a
macroevolutionary scale, since we cannot do experiments with macroevolution.
(2) Perhaps for this reason many authors (particularly population geneticists)
are suspicious of tendencies which are not so anchored. Without a
microevolutionary basis, the fear is that such constraints may fail either to
have an explanation or to be acceptable to many authors. (3) Without a basis
in microevolutionary mechanisms, there is no guarantee that the tendencies
advanced are even biologically or physically possible. Clearly, this is a
requirement for any evolutionary constraint. (4) If macroevolution is a product
in whole or in part of universal or nearly universal microevolutionary
processes, microevolutionary mechanisms which yield constraints are likely to
have broader effects or at least more predictable ones in macroevolutionary
change. This is a feeling at least as old as the nineteenth century
uniformitarianism of Lyell and Darwin. Darwin after all did not give us the
idea of evolution -- rather he provided microevolutionary mechanisms which
made it both possible and inevitable.

It may be too strong to say that any evolutionary constraint must have a
microevolutionary basis, since some processes might operate on too long a time
scale to count as microevolutionary. But any such constraint must at least
meet the third condition -- consistency with microevolutionary mechanisms.
Nonetheless, the obvious advantages of a foundation in microevolutionary
processes, described in the preceding points, make it a strong desideratum if
such a constraint can be demonstrated to have such a basis.

There are already two widely recognized constraints on the evolution of
adaptations that are microevolutionary in character. The first, actually three
conditions, is what Lewontin (1970) has called "Darwin's principles." These
conditions, appropriately summarized by saying that for evolution to occur
there must exist heritable variance in fitness, are advanced as separately
necessary and jointly sufficient conditions for evolution. They represent the
central assumptions of the argument for evolution by natural selection.

A second constraint, clearly articulated by Lewontin (1978), is what he
calls the principle of quasi-independence.[5] This states that it must be possible
to change and select for a given trait without at the same time changing a
large number of other traits of the phenotype. This condition is necessary to
secure local adaptation by selective "fine tuning" to locally variable conditions,

and it allows for "piecemeal engineering" of the phenotype. Perhaps more importantly, if genetic changes which increased fitness along one dimension were all highly pleiotropic (and these changes were not adaptively correlated, as in allometric growth), the probability that the net effects of that change would be adaptive is exceedingly small, and each evolutionary change would have to await the occurrence of a "hopeful monster." Alternatively, if there were variance in the average degrees of pleiotropy among phenotypes, those with less average pleiotropy would respond more quickly to selection than those with more, and the net effect should be selection for increases in the "decomposability" of the phenotype (Wimsatt 1981). Presumably, the cumulative evolution, in multiple cases, of modifier loci to decouple the "main effects" of a gene from its various side-effects, could have this result. No one has, to our knowledge, investigated this question.

We wish to argue that there is a third constraint which, unlike the preceding, applies specifically to the evolution of complex adaptations. It is exceedingly robust, and is plausibly met for all known complex adaptations of all organisms. This constraint is closely related to another one which limits the number of genes that can be maintained by selection, and which has been discussed in the debates over genetic load. There are possible ways of avoiding this latter constraint, or of attenuating its effects. These means of escape provide what may be thought of as "adaptation pumps" which, over extended periods of evolutionary time, should increase the number of genes that can be maintained by selection, and thus the complexity of adaptations which they may produce. These constraints are best explained through discussion of simulations which we undertook to test the robustness of conclusions reached by Stuart Kauffman on the evolution of complex gene control networks.

A Simulation Result

Kauffman for the last eighteen years has been pursuing a research program of modelling the behavior of gene control networks as networks of Boolean automata, composed of elements which receive and send "on-off" signals to one another and whose states (and the signals they send) are determined by the combination of their inputs (Kauffman 1969, 1985, 1986). He has shown that networks which are randomly constructed subject to certain constraints (in this case, local conditions which serve to define the problem) display a large variety of biologically interesting properties. These constraints are conditions such as the number of circuit elements, the number of inputs to each element, and the Boolean functions which determine how the output of the element depends on the combination of its inputs. The circuits are randomly constructed in terms of what is connected to what and the starting states of the circuit.

Kauffman's interest in ensembles of randomly constructed networks derives from the conviction that the number of control genes in metazoans is too large for selection to maintain specific "wiring diagrams" for such networks in all of their detail. He thus believes that at least a large fraction (presumably the overwhelming majority of such connections in large networks) must be established and changed through drift processes. It then becomes interesting to ask whether there are any properties which are characteristic of virtually all members of an ensemble of such randomly constructed networks, since these properties should be realized in all or virtually all complex organisms. Kauffman calls these properties, of which he has found a significant number, generic, and he regards their presence (which is statistically inevitable, or nearly so) as generic constraints on the architecture and behavior of such networks.

This modelling strategy and the search for statistically likely properties was inspired by statistical mechanics. It also motivated Frank Rosenblatt's (1962) "perceptron" models, the ancestor of the recent cognitive and neurophysiological "parallel distributed processing" network models. Rosenblatt's worries about detailed neural network models exactly paralleled Kauffman's worries about gene control networks. He argued that the genome did not have sufficient information to specify all of the connections in a complex neural network, and that in any case the structural variability from network to network in different individuals of the same species had to be consistent with their convergent and species typical behavior. Both of these forced the search for generic properties (which Rosenblatt somewhat misleadingly called "genotypic"). This distinctive modelling strategy used by Kauffman and Rosenblatt is very powerful, (see also Burian 1986) and is worth further attention. It is, of necessity, the strategy of choice for modelling systems that are too complex to suppose that selective processes can fine-tune all aspects of their structure and behavior.

Such statistically likely properties (which are in effect high entropy properties of the ensemble) are of interest for several reasons: (1) They are likely properties of any network whose structure is, to a significant extent, a product of drift or other random processes. (2) They are likely to have a high degree of heritability (or for neural networks, functional redundancy and plasticity), even in the face of substantial mutation, because virtually all other circuits which are mutational products of the circuit in question will also exhibit the properties. Heritability is (from Darwin's principles, discussed above) a *sine qua non* of any evolutionarily significant properties. (3) These properties, will for the same reason, appear to be strongly canalized or homeostatic, being resistant not only to mutations, but also to a variety of nongenetic (environmental or somatic) perturbations. (4) These properties will not require the invocation of selection for their explanation. One does not have to select for them if one gets them "for free."

All of Kauffman's earlier work was connected with the discovery and analysis of such properties, but recently he has undertaken simulations to establish what he had earlier taken for granted -- that selection cannot in fact maintain the detailed structure of large gene control networks. It is this work that is particularly relevant here. In these simulations, Kauffman used a simpler model of the structure of gene control networks than in his earlier dynamical studies with Boolean networks. We have recently begun testing these conclusions in models similar to his original ones, and our preliminary results appear to qualitatively confirm those discussed below.

In this simpler model, the networks are modelled as directed graphs, with genes being nodes and arrows being influences of one gene by another (fig. 1). Mutations in these networks are modelled as the random reassignment of the head or tail of an arrow, so that with n connections, there are 2n mutable "sites." The number of genes thus does not affect the number of mutable sites in these simple models, though it does determine the back-mutation rate, since the number of genes determines the number of alternative possible states of a connection, and with fewer possible states to mutate to, the probability of back-mutation increases. The ratio of the number of genes to the number of

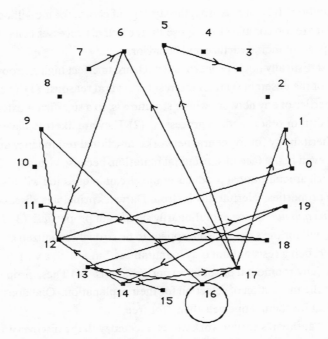

Figure 1. A directed graph representation of a gene control network with 20 genes and 20 connections produced by 1000 iterations of mutation from a closed loop model gene system of 20 genes and 20 connections. This net is indistinguishable in its generic properties from a net which is constructed at random.

connections is also significant because it determines the connection density of the network, and through this, a number of its generic topological properties.

In his simulations, Kauffman started with a population of "ideal" networks, all of which had a certain (arbitrarily chosen) specific circuit structure, subjected them to mutation, and allowed them to reproduce with a fitness determined by the proportion of connections they shared with the "ideal" network. (See Kauffman 1985, or Schank and Wimsatt in press, for further discussion.) In a population of 100 circuits (each with 20 genes and 20 connections, a mutation rate of .005 per mutable site per generation, and a selection coefficient of .05 per connection), he found that after 1000 generations, even in the presence of fairly strong selection (though with very high mutation rates!), approximately half of the "good" connections were lost

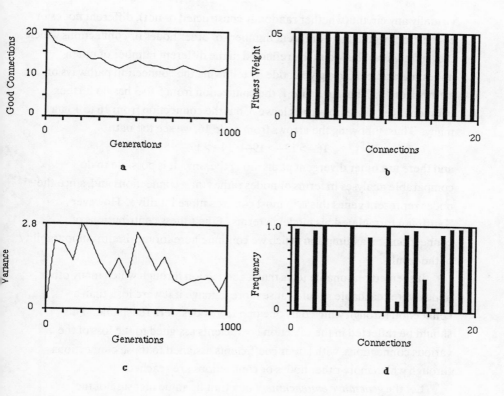

Figure 2. A sample simulation of the evolution of a population of 100 model gene networks with 20 genes and 20 connections over 1000 generations. For explanation see text.

at random (fig. 2). This loss was sufficient to make the relevant properties of networks in the population generic even if they were not at the start. (The circuit described in figure 1 provides a specific example of the outcome of this kind of transformation.) Kauffman concluded from this, as well as from other simulations and arguments, that selection could not maintain specific wiring diagrams in large circuits, and that the vast majority of relevant properties of large gene control networks were generic, and not explained through the operation of selection.

We were bothered by one assumption in Kauffman's simulations, namely the assignment of equal fitness decrements for the loss of any connection. In virtually any circuit (whether randomly constructed or not), different nodes or connections would have different numbers of other nodes or connections which they influence. This would be reflected in the different number of other nodes reachable from a given node by following the connection pathways of arrows from it. Thus, in figure 1, the connection from 5 to 3 has no further consequences, since no arrows leave 3, but the connection from 16 to 13 has many. Thus, following the arrows from node 16, we see the path

$$16\text{--}>13\text{--}>19\text{--}>14\text{--}>17\text{--}>5\text{--}>3,$$

and there are other divergent paths along the way. It is possible to do comparable analyses in terms of nodes rather than connections and, since the nodes represent genes this is in most contexts more intuitive. However Kauffman formulated his model in terms of the fitness contributions of connections, an assumption which we continue here for maximum comparability of the results.

It seemed reasonable to us that a connection through which many other nodes were reachable should cause more damage if it were lost, than a connection through which only one or a few nodes were reachable. This should be reflected in the selection coefficients assigned to the loss of the various connections, with larger coefficients assigned to those connections through which more other nodes or connections are reached.

Let the *generative entrenchment* of a trait be some measure of the number of other traits affected by changes in that trait. We have argued elsewhere (Wimsatt 1986; Schank and Wimsatt in press) that traits with greater degrees of generative entrenchment, such as traits earlier in development, should be more conservative in evolution. Rasmussen (1987) has found convincing confirmation of this hypothesis in a review of experimental studies on developmental mutants in *Drosophila*. Traits with greater degrees of generative entrenchment should thus be assigned greater selection coefficients favoring their preservation. We have tried a number of different measures of generative entrenchment (Schank and Wimsatt in press), and the results we describe are robust over these different measures. It should be apparent that *the presence of different degrees of generative entrenchment for the different nodes or connections in the network is a generic property of such networks.* This is a fact that Kauffman notes in passing (Kauffman 1985), but he does not use it in calculating fitnesses. Thus our models are, by his own

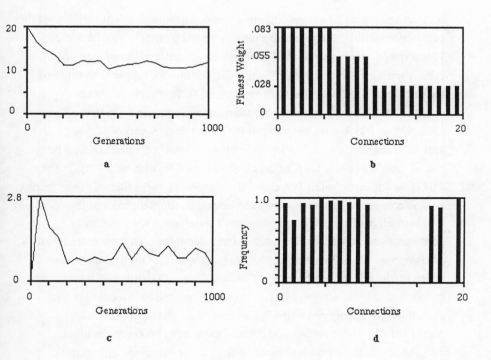

Figure 3. A sample simulation of the evolution of a population of 100 model gene networks with 20 genes and 20 connections over 1000 generations under the same conditions as in figure 2, except that fitness decrements were made proportional to the generative entrenchment of the connections and normalized so that they summed to 1. The baseline fitness (fitness contribution not affected by mutation) per connection was 0. The "good" connection equilibrium, as in the preceding simulation appears to be about 10.5 out of 20 connections, but the remaining connections are strongly biased toward the more generatively entrenched ones.

assumptions more realistic. We have now done simulations paralleling those of Kauffman in different sized populations of circuits with a variety of different sizes, connection densities, mutation rates, and with a variety of different fitness functions reflecting different measures of generative entrenchment, all with roughly the same results, a sample of which are presented in figure 3.

In this simulation, as in others, we observed the same decline in the proportion of good connections to an equilibrium level of about 50% as were observed by Kauffman (fig. 3a). However, unlike his case, in which good connections were lost at random, the losses occurred almost entirely among the connections with lower generative entrenchment (compare figs. 3b and 3d). This differential preservation of generatively entrenched connections occurred even when their selection coefficients were not much stronger than those of the ones lost. Kauffman used a fitness loss of .05 for mutations in each connection. The range of fitness decrements per connection in figure 3 are from .028 to .083, but no losses occur for fitness decrements of .055 or greater. In figure 4, with smaller differences in selection intensities, all but one of the losses occur for fitness decrements of .038, with only 1 loss in the .051 class and none in the .064 class. By themselves, these figures are perhaps not too convincing (given that they are stochastic simulations), but they are representative of a much larger number of simulations for a variety of parameter values with similar results. Even deeply generatively entrenched connections may be lost, but they are lost at lower rates than connections which are less deeply entrenched. The net effect is that although a large number of connections were lost even though favored by selection, in each case, the most deeply generatively entrenched connections tend to be preserved--through runs in some cases of as many as 5000 generations.

Although these simulations were designed to assess the effects of differential generative entrenchment of phenotypic traits, this assumption was used only for the assignment of different selection coefficients. The assumption of differential generative entrenchment was not used in any other way in the simulations. *What the simulations thus suggest is that even relatively small differences among selection coefficients of genes (for whatever reason--whether due to differential generative entrenchment or just to their differential adaptive importance) are sufficient to bias the preservation towards the more strongly selected ones, even in the face of quite strong losses due to drift forces.*

These results have been further confirmed in recent simulations with a new and more realistic model. The significant changes in this model are: (1) For computational simplicity, the fitness decrements are assigned to connections, rather than being calculated from their generative entrenchments. As noted above, the dynamics of the model are affected only by the resultant selection coefficients, not by the generative entrenchments. Plausible distributions of selection intensities and the frequencies with which they are

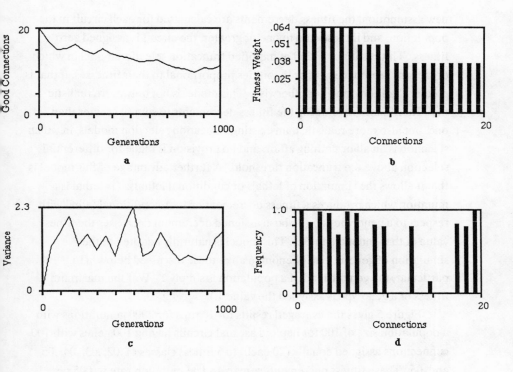

Figure 4. A sample simulation of 100 model gene networks with 20 genes and 20 connections over 1000 generations. See text for explanation.

assigned to connections for randomly constructed networks can be determined, and then plugged into this model, but the model also allows arbitrary manipulation of the distribution of selection coefficients. (2) Kauffman assumes that the fitness contributions of all of the connections sum to one. With a concept of relative fitness, it is plausible to regard the net contributions of all genes as summing to this quantity, but implicit in his model is the assumption that the sum of fitness *decrements* for deleterious mutations in all genes is also one. This is implausible. It would require, for example, at most one lethal mutation, and if there is one, that all other mutations are neutral! In the new model, it is assumed that the sum of fitness decrements can (and in general will) be greater than one. In employing this new assumption, the fitness decrements are calculated for each circuit in the population, and if these sum to one or greater, the circuit is assigned zero fitness. This results in a kind of modified truncation selection model in which organisms reproduce with probabilities proportional to their fitnesses, if that is greater than zero, and not otherwise. This model is not only more realistic than Kauffman's in allowing the fitness decrements to sum to greater than one, but also more realistic than existing truncation selection models, in which selection is an all or nothing affair, and no provision is made for differential selection above the truncation threshold. A further advantage of this model is that it allows the simulation of lethals or conditional lethals. (A lethal is a mutation which produces a fitness decrement of 1. A conditional lethal with respect to a genotype is one whose assigned decrement decreases the fitness value of that genotype to 0.) This model is naturally adapted to the simulation of genetic load phenomena like those discussed below. (In particular, the genetic load of a population is simply 1 - W if the maximum fitness of a genotype is assigned the value 1.)

Figure 5 gives the averaged results for 10 runs for 2000 generations with a population size of 100 for haploid asexual circuits having 100 genes with 100 connections assigned equally (20 each) to 5 fitness classes of .02, .03, .04, .05, and .06. These fitness decrements sum to 4. The mutation rate is .005 per connection end per generation, yielding a mean mutation rate of 1 per organism per generation. Because of the larger number of genes, the back mutation rate is of the order of 25 times smaller than for Kauffman's simulations, enough so that connection losses are effectively irreversible over the time of the simulation once the mutation becomes fixed. The mean population fitness (plotted every 500 generations as a percentage of maximum fitness) fluctuates at around 10-11%, indicating that a reproductive potential of 10 or more is required for the population to maintain itself.

The top graph is a stacked bar graph of the cumulative frequency of good connections in all 5 fitness classes. This is analogous to the frequency graphs in figures 2a, 3a, and 4a, but includes also some of the information given in figures 2d, 3d, and 4d. This frequency stabilizes at about 70%, significantly higher than Kauffman's (in spite of a circuit size 5 times as great) because the total fitness decrement is now 4 rather than 1, and circuits which lose even a few more connections are lethals. The grey scale represents

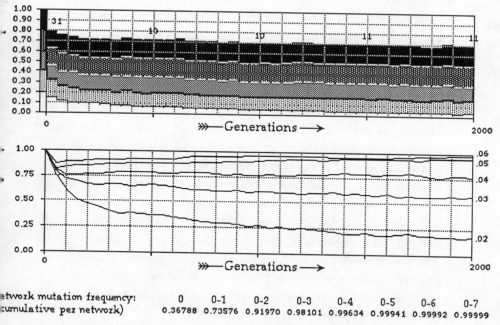

Figure 5. Evolution of frequencies for connections in 5 fitness classes in gene control networks having 100 genes with 100 connections. For explanation see text.

the relative selection intensity of the various classes, with black being the class with the largest fitness decrement. In the bottom graph, the frequencies of the various classes are plotted as proportions of their initial frequencies.

There are two particularly salient results of these simulations for present purposes: (1) As with the simulations for smaller circuits, there is a strong differential bias favoring retention of the more strongly selected alleles. This differential bias persists in the average of multiple runs, and is illustrated in both the top and bottom graphs. (2) A surprising phenomenon emerges in the bottom graph -- a bifurcation between the fates of the connections with lower and higher fitness decrements. The two lowest fitness classes (.02 and .03) decrease in frequency throughout the course of the simulation, but the two highest fitness classes (.05 and .06) initially decrease but then increase after a minimum reached in generation 50. (The frequencies are plotted only every 50 generations.) The middle fitness class (.04) equilibrates after about generation

100 at 75-80%. This bifurcation process provides a possible explanation for the significantly different fates of connections with relatively small differences in their selection coefficients.

But why does this bifurcation process occur? The answer is quite surprising, and has broader implications for the interpretation of selection models. Suppose that we start, as Kauffman did, with a population of identical organisms, each with an ideal circuit having 20 connections, a fitness of 1.0, and a fitness decrement of .05 per connection for any mutated connections. A mutation to one of these circuits gives it a 5% loss in relative fitness. Now suppose that, after a number of generations under the pressure of drift and random mutation, the fittest circuit has but 10 good connections, and a fitness (on the original scale) of .50. If it now loses a connection, it loses, as before, .05 in fitness, but this now represents a 10% relative fitness loss. If it had but two good connections, the loss in relative fitness would be 50%! This is a surprising result, because *it suggests a kind of frequency dependent selection occurring in what was supposedly a strictly additive model with constant fitnesses! Indeed, it raises new questions about whether it is even possible to have frequency independent selection.*

This frequency dependent selection does not require a truncation selection model, and holds even for Kauffman's model, although in general truncation selection models (of the type studied here) result in lower mean population fitness, and thus generate a larger effect. It occurs also for multiplicative fitness models with truncation selection. This frequency dependence occurs because the relative fitness loss is given by the size of the fitness decrement *relative to the fittest genotype actually found in the population*, not relative to the fittest genotype at the start, or the fittest genotype that could be constructed with the genes in the gene pool of the population. It should become a significant factor in proportion to the decline in mean population fitness relative to its maximum possible fitness -- in other words, in proportion to the genetic load of the population. Deterministic population genetic models in which any possible genotype occurs with some finite frequency (because of the implicit assumption of an infinite population size) do not produce this dependency, because the fittest genotype in the population does not change over time, but stochastic models with finite population sizes do.

The bifurcation process requires more than this kind of frequency dependent selection -- it requires also the presence of different fitness classes. What happens, we suggest, is that starting from an initial highly fit population, mutations and drift cause the loss of alleles in all classes, although at a higher rate in the lower fitness classes. As the mean fitness of the population and the value of the fittest genotype in it decrease over time, the relative selection coefficients of alleles in all fitness classes increase by the same scalar factor (of $1/W_{max}$) until the fitter ones stop decreasing, reaching a rough equilibrium.[6] The alleles in the lower fitness classes have not yet reached their rough equilibrium, and continue to be lost, further increasing the selection coefficients of the alleles in the higher fitness classes, which, as a result, increase in frequency. In effect, the population reaches a rough fitness equilibrium (in fig. 5, from roughly generation 500 on), *after which evolution proceeds by the substitution of fitter alleles for less fit ones at other loci*! Further work is required to determine the magnitude of the effects of this proposed mechanism under different conditions.

The different results obtained from models with uniform versus heterogeneous selection coefficients have quite an interesting further consequence -- a constraint on the evolution of complex adaptations.

Generalizing this Result: A New Evolutionary Constraint

Kauffman's results are actually to be expected in the light of the theory of genetic drift and the neutralist-selectionist debates. They appear different from most of those discussions (Wallace 1968; Crow and Kimura 1970; Lewontin 1974) because he is not looking at drift among neutral alleles, but at loss of selected alleles. This topic was discussed by Sewall Wright (1931), who derived equilibrium distributions of gene frequencies as a function of population size, mutation rates, and selection intensities (Wright 1931, fig. 19, p. 148), and suggested that cases with heterogeneous selection coefficients can be treated as aggregates of his various distributions. Wright's parameters have somewhat different values, his curves are derived assuming sexual rather than asexual organisms, and there is no back mutation, but his figure shows relationships between selection intensity and probability of loss of alleles that are qualitatively similar to our results.

It seems clear both from Kauffman's simulations and from the various genetic load arguments that for given population sizes and mutation rates, selection cannot maintain too many genes at one time, and that if they all have the same selection coefficients favoring their preservation, they will be lost at random. That they will be lost at all requires finite population size and invocation of the theory of genetic drift. The second part of the statement requires only a symmetry argument: *As far as the theory is concerned, two alleles are treated equally if they have the same selection coefficients. Thus if a class of alleles all have the same selection coefficients, and some of them are to be lost, there is nothing to favor any preferred subset of them, and they will be lost at random.*

If a significant number of genes are being selected for, those that will be preserved will be those with the higher selection coefficients. Consequently, *the genes contributing to a complex adaptation of the phenotype or unit of selection must show differences in the size of their fitness contributions -- even if these contributions are all positive -- if they are to be saved in any systematic fashion. If we suppose that for a complex adaptation to function at all, some subset of the genes coding for it are essential, then it is a requirement that there be differences in their selection coefficients.* This is the first of the two constraints we will discuss.

This principle applies when the number of genes (or other hereditary units) contributing to an adaptation is large enough so that their average selection coefficient is relatively small -- small enough that mutation and genetic drift become significant causes of their loss over the relevant time scale. In these circumstances, if the genes all make equal contributions to fitness, they will be lost at random (with equal probability and at equal rates), and different ones will be lost in different descendants. *This rules out the possibility of a common core of genes (and traits which they determine) which can act as foci or bases for the development of more complex adaptations.*

This constraint might seem to follow trivially from Darwin's principles if the gene is taken as the unit of selection, since then Darwin's principles require that these genes must show heritable variance in fitness. It is important to see that this is a different and additional constraint. It is not just that the genes in question must show heritable selective superiority to the mutants which are being selected against: this is what is required by Darwin's principles, and this much is included in Kauffman's original model, with results depicted in figure 2. More is required -- that they must show heritable *differences* in their selective superiority relative to these mutants if a subset of them is to be reliably preserved, and the subsets that will be differentially preserved will be those with the larger selection coefficients. This is the result depicted in figures 3 through 5.

Moreover, there is no reason to believe that this principle must be restricted in application to the genic level. The logic of the argument suggests that it should apply to gene complexes or to any other higher level heritable units which together act to produce a complex adaptation. To put it more generally: *for the evolution of complex adaptations in units of selection at a given level, there must be differences in the size of contributions of hereditary units at some lower level of organization to the fitness of those higher level units*. In other words, it applies to the evolution of complex adaptations at any level of organization *whether or not* individual genes are taken to be the units of selection. A hierarchical model of multiple genetic units of selection, consonant with the theoretical work of Wade and other is presented in Wimsatt 1981. The principle could apply at several levels simultaneously if there are complex adaptations of units of selection at each of these levels.

This constraint is anchored in two independent ways in the nature of physical and biological phenomena:

The improbability of structures which violate this constraint. First, it is almost impossible to build any interesting (i.e., differentiated) complex structures in which *all* components or their determinants make equal contributions to fitness. Complex artifacts can fail in a number of ways, but only rarely would it be true that different failures have equally serious effects. (Consider the range of defects, from the superficially cosmetic to the mechanically fundamental in most new cars.) The wide scope of this claim can be seen by looking at two classes of *prima facie* counterexamples, and understanding why they fail.

(A) A strictly serial structure, in which any mutation is lethal and thus equally disadvantageous is possible -- like the Christmas tree lights in which failure of a single bulb extinguishes the string. But such structures are rare and seldom large in real organisms, though probably for a different and even stronger reason than this constraint -- a fact which is capable of exerting a local selective influence in microevolutionary time: The common existence of parallel pathways or other modes of functional redundancy is presumably selected for because large serial structures (structures which fail if any of their components fail) are notoriously unreliable.

Thus, if we have a serial string of n paths or components, each of which has a probability of d of failing, and failures are independent of each other, the probability of successful operation for each component is (n - d), and the probability of successful operation for the whole string is $(1 - d)^n$, since all the components must work if the serial string is to work. No matter how small d is, as n gets larger (i.e., as the serial structure gets longer), the quantity $(1 - d)^n$ approaches 0. In this situation, the introduction of parallel paths is an effective way of increasing reliability, since parallel networks fail only if all the paths fail. With a failure rate of d per path, and n parallel paths, the failure rate for the structure is d^n, which is always less than d, so the introduction of parallels increases reliability. But random placement of parallel "shunting" paths along parts of the string (as might be expected with random mutations) will generate inhomogeneous distributions of selection coefficients on the elements of the paths.

This argument does not require that mutations in the serial structure be lethal, but only that the overall structure is sufficiently important that mutations which increase its reliability will be selected. In this context, the evolutionary addition of parallel shunts is to be expected. Thus selection on a microevolutionary time scale would tend to transform cases in which a large number of components, or the genes coding for them, have equal selection coefficients into cases where they are assigned unequal selection coefficients. Indeed, if mutations are capable of adding links to the network, randomly placed mutations would have the same effect, whether or not they are selected. The topology of real examples of biochemical networks clearly suggests that the latter case -- with unequal selection coefficients -- is the rule rather than the exception.

(B) It is also possible with strictly parallel redundancy of similar components to get large adaptive structures whose parts contribute equally to fitness. Large adaptive or genomic structures whose parts contribute equally to fitness are generally cases in which any combination of the components or genes will do, as long as enough of them are functional. Then mutational losses of particular genes (or components) are irrelevant. In this case, if we have m components, each with independent failure rates of d, and at least k of them must work for the system to work, the probability that the system will work is given by:

$$1 - \sum_{i=m-k}^{m} C(i,m) (1-d)^{m-i} d^{i}$$

Here the sum gives the probability that from m-k to m of the components will fail, causing the system to fail, and 1 minus this sum is the probability that k or more of the components will work, and thus that the system will work. C(i,m) is the number of combinations of m things taken i at a time, and is the number of different ways in which i components can fail if failures occur independently of one another. To a first approximation, this equation characterizes cases like that of the multiple copies (m) of DNA which code for ribosomal RNA, where multiple copies are required to generate enough ribosomes to maintain sufficient capacity (k) for protein synthesis. In this model, though not in real life, "sufficient capacity" is an all-or-nothing affair, and a more realistic model would recognize this. Presumably the number of copies present shows some excess over this minimum capacity to allow for mutational losses. Mutational losses (d) do not denote the per locus mutation rate but the total proportion of malfunctions, including not only the total number of genetic mutations but also the sum of somatic mutations and other malfunctions in the expression of those genes in the cell.

This represents a genuine exception to the requirement for an inhomogeneous distribution of selection coefficients. It does so by providing a case in which no particular subset, but only a given proportion (k or more out of m), of genes is required for proper functioning of the functional system or adaptation. On the other hand, not just any case of redundancy or increase in capacity secured through the duplication of like components fits here. What is required is that the redundancy be at the relevant level, in this case the genetic level. Two lungs, eyes, gills, or kidneys in bilaterally symmetric organisms are presumably the result of the operation of the same set of genes, albeit in different cells, and if there are no redundant copies of them, this model does not apply. Similar comments apply for many other cases of component redundancy -- to the number of liver cells found in normal organisms, or for the cells of any organs in which the productive capacity of the organ is a roughly linear or even just a monotonically increasing function of the number of functioning cells in the organ. In these cases, the above model applies (with appropriate qualifications) to the calculation of reliability of the compound organ, and the parameters m and k are presumably subject to selection, but the model does not apply to the analysis of the reliability of the structure of control genes and structural proteins which produce the cell types in question, since there may be only one copy per cell of the genes which produce the differentiated cell type. Thus it seems likely that strictly parallel modes of organization in networks will not present frequent counterexamples to this thesis.

There is perhaps only one class of adaptations which could escape this constraint. If an adaptation were generic in Kauffman's sense, then almost any structural rearrangement of the gene control network which produces it would still yield a structure with the property in question. Such structures would escape the constraint by a statistical inevitability which would be a kind of massive statistical redundancy --they would be effectively immune to mutation. Some of these cases may behave qualitatively or approximately like the preceding "k-out-of-m" case, but there is no obvious reason why they must. Kauffman has found such properties in his modelling efforts, and if he is right, they are a very important subset of adaptive properties. Nonetheless, they represent a very small fraction of the total of known adaptive properties of organisms, and the rest seem all too sensitive to degradational mutations, as is readily demonstrated by the phenomenon of inbreeding depression.

In sum then, it is highly improbable that adaptive structures should have equal contributions to fitness from their various parts, and circuits or parts of circuits which have this property will tend to lose it through selection for increasing reliability.

The long-range evolutionary instability of structures which violate this constraint. Genes which make larger contributions to fitness will, for given mutation rates and given effective population size, have a longer expected lifetime before they are lost through drift. In systems with inhomogeneous distributions of selection coefficients, the bifurcation phenomenon discussed in the last section can be expected to amplify this effect. The presence of recurrent and back mutations does not really change this picture, for then we have an equilibrium distribution of frequencies for alleles at different loci, and those genes with lower selection coefficients spend more time at low or zero frequencies. This is demonstrated in our simulations and predicted by theory (Wright 1931).

There is a very important further consequence of these differences in expected lifetime. Evolution is opportunistic -- it uses what is already there, supplemented with new additions to modify existing or to make new adaptive systems. Assume that there is a given probability per unit time that a gene will acquire a dependent adaptation -- a mutation at another locus (e.g., a modifier gene, though not all such genes would normally be thought of as modifier genes) which is adaptive in the presence of that gene, but only in its presence. The acquisition of adaptations of any sort is assumed to be a relatively slow process, so this probability is assumed to be small, and to occur at a significant rate only over extended periods of time. Assume for simplicity that this probability is the same for all genes. Then genes which stay around longer have a higher probability of acquiring such adaptations, and a higher expected number of such adaptations that they will acquire. Such genes now have greater generative entrenchment, and degradational mutations in them will cause greater fitness losses than would have been the case before they acquired their dependent adaptations.

If genes with larger net fitness effects stay around longer, and genes which stay around longer accumulate more generative entrenchment than genes with smaller fitness effects, the net effect will *amplify* initially small differences in the genes' contributions to fitness, with important genes (making larger contributions to fitness) tending to become more important, and less important genes being lost due to drift. *Thus selection should act over long evolutionary periods as a symmetry-breaking force, so that systems with equal or nearly equal selection coefficients for their parts will be driven by selection to greater variability in the distribution of the parts' selection coefficients.* Even if we start with equal selection coefficients, stochastic fluctuations due to drift processes and the random acquisition of dependent adaptations by different genes should generate the variability necessary to prime this symmetry-breaking force.

For one important class of genes (particularly important, since this class enters in two of the mechanisms to be discussed below) this argument can be made even stronger. In the discussion of the simulations, we compared the results when all the connections or genes had the same selection coefficient (Kauffman's case) with the results when there were differences in the selection coefficients. For those purposes, it only mattered what the selection coefficients were, and not the reasons for the differences. In particular, it did not matter whether large selection coefficients arose because the gene or connection was deeply generatively entrenched, or whether it was relatively unentrenched but simply very important for survival. There is, however, another dimension to the problem.

Generatively entrenched genes will, by definition, have more genetically mediated consequences downstream of them. They, therefore, provide more opportunity for the accumulation of modifier genes which act on various of these consequences. This would be true even if the sequence of developmental consequences were linear, but it is more like a branching tree. Wallace Arthur (1984) has argued that there should be a significant increase in the number of genes acting at later stages in development, in large part because there is an increase in the number of differentiated subsystems through the course of development, each of which is presumably fine-tuned by selection through the differential action of a number of genes. While it is impossible at present to attach numbers to or even to demonstrate such an increase, this assumption is highly plausible.

If this is true, then genes which have a large selection coefficient *because they are generatively entrenched* are different from genes that have the same selection coefficient but less generative entrenchment; the difference is that there should be more ways in which members of the former class can acquire potential modifiers. Coupled with the argument that genes which stay around longer should acquire more modifiers is the implication that *the class of genes that should acquire modifiers at the greatest rate and in the largest numbers should be those with large selection coefficients achieved, because they are deeply generatively entrenched.* Thus, not only should selection favor inhomogeneous distributions of selection coefficients, but it should do so in a way which particularly favors genes that are generatively entrenched. We will return to this point later.

Thus, this constraint, denying equal fitness contributions by the parts of an adaptation or the genes generating it, is exceedingly robust. It is guaranteed both by the improbability of the occurrence or construction of such systems, and by selection away from it. Selection acts in the shorter run to favor adaptations (parallel "shunts") that increase the reliability of serial structures, but it, or random mutation, would also produce inhomogeneities among the selection coefficients. It acts in the longer run through the differential acquisition of dependent adaptations by genes with larger selection coefficients, and particularly by those genes whose selection coefficients are large because of substantial generative entrenchment. This also acts to amplify variability in the distribution of selection coefficients associated with genes that produce parts of these structures.

We do not have information on the distribution of selection coefficients for mutations in all parts of any organism. However, any organism studied has shown a strong heterogeneity in the severity of its mutations. Ever since Darwin, and well before Mendel -- it was noted by Koelreuter in the eighteenth century (Olby 1966) -- this heterogeneity (in the nineteenth century) of variations rather than of mutations has seemed too obvious to mention, which is probably why it went unnoticed.

We have argued that it is important for the evolution of complex adaptations that there be a heterogeneous distribution of selection coefficients for the genes that generate it. For a variety of reasons, this requirement is met, essentially universally, in nature. This requirement places a constraint on the evolution of complex adaptations *only* if it is not possible to maintain an arbitrarily large number of genes with equal selection coefficients for significant periods of time, for only then do we have to worry about *which* genes are maintained by selection. The discussions by Kauffman and ourselves about genetic load, and the simulations suggest that there is such a limitation. The impossibility of doing so is the second important constraint on the evolution of complex adaptations.

Simulation and theoretical work by Kauffman (1985) suggests that the maximum number of genes making equal contributions to fitness which can be maintained for intermediate periods of time (of the order of 1000 generations, the length of Kauffman's simulations) may be as few as 20 (in haploid asexual organisms) for the mutation rates he considered. While our own (as yet unpublished) work suggests that the number is somewhat larger with a truncation selection model or for diploid sexual organisms, this number still seems likely to be disturbingly small. (See also Kauffman 1985 for relevant theoretical discussion.)

This remains true even if we substitute more reasonable mutation rates as long as we are trying to maintain perfect or near-perfect "wiring diagrams" in large genomic structures. This is the most serious unrealistic assumption of Kauffman's model, and the robustness of these conclusions need to be tested for genomic structures in which redundancies in the circuit mean that no significant fitness cost is imposed unless the deviations from the "ideal type" are greater than a certain amount. One would expect that with a selection model that took account of these redundancies, (i.e., another variety of "truncation selection" model than is considered here), it would be possible to maintain a larger number of genes through selection. Indeed, diploidy represents one such mechanism of increasing redundancy, and it is revealing that our preliminary simulations of diploid circuits with dominance suggest that is it possible to maintain a somewhat (but not enormously) larger number of genes. There are other modes of redundancy than diploidy, however, and these possibilities deserve to be investigated. In any case, it is plausible to say that the same problem would reappear for a larger genome size. Kauffman (1985) estimates that the genetic control structure of large metazoan genomes is of the order of 100,000 genes. If this estimate is anywhere near the correct answer, it seems unlikely that this change alone would solve the problem.

For this reason, it is desirable (if we are to defend a selectionist, rather than a neutralist account of evolution) to find some way around this constraint -- some way in which a much larger number of genes can be maintained by selection. A number of such ways have been discussed in the literature surrounding the neutralist- selectionist controversy. We propose four ways, discussed in the next section, by which the number of genes so maintained could be increased by several orders of magnitude over extended periods of time. Two of them (the first and last) have already been suggested in the literature, but two of them are, to our knowledge, new at least to the recent literature. If any of these mechanisms has a significant effect, there should be a clear sense in which it would be appropriate to speak of progress in evolution.

Four Possible Mechanisms for
Avoiding the Limitations of this Constraint

We can delineate four possible mechanisms using this constraint and other generic features of complex phenotypes and relevant selection regimes.[7] These would generally have the effect of increasing the number of genes available to determine a complex adaptation which can be maintained in the face of the degradational forces of mutation, recombination, and drift. We will make reference to theories of genetic load, which are not without problems (Wallace 1968; Lewontin 1974), but the conclusions drawn are not particularly sensitive to the problematic assumptions of these theories. We will focus on the number of genes that can be maintained at or near fixation, and thus on the maintenance of the "constant" portions of the genome. Classical discussions of the problem of genetic load have, by contrast, focused on the maintenance of genetic variability -- that is, on the "variable" portions of the genome. The two problems are related: if a constant locus can be maintained at lower cost in genetic load due to changes in the mutation rate or the time of expression of the trait that it affects, or if a variable locus, through transition to a constant locus, can be maintained at lower cost, this leaves more of the allowable load for the maintenance of genetic variation at the variable loci. Thus load reductions in either area contribute to the solution of problems in the other.

This supposes that a type of organism is characterized by a maximum reproductive potential, R, the number of offspring it can have under the best possible conditions. This reference state is defined as that state in which a population is composed entirely of individuals having the fittest genotype for that kind of organism. The genetic load, L, induced by a condition, mechanism, or constraint may be defined as the proportion of this reproductive potential which is removed by that condition.[8] Thus, if the genes in a genotype mutate to less fit states in some offspring, the offspring population has lower fitness than the parent population, which is reflected in a lower reproductive potential. Other mechanisms and forces also reduce the fitness of the population. Thus, if the fittest genotype is heterozygous at a given locus, segregation will produce less fit homozygotes among the offspring, reducing the mean population fitness. Similarly, recombination among the homologous chromosomes of an epistatically superior genotype will produce less fit offspring in which the adaptive gene complex is broken up. Crow and Kimura (1970) discuss a variety of different types of genetic load.

The load induced by a given mechanism is assumed to grow cumulatively (sometimes additively, sometimes multiplicatively, depending on the mechanism) with the number of loci it affects, and the loads induced by different mechanisms are added together to get the net genetic load on organisms of that type. If the product of the net load with the reproductive potential, LR, is less than 1, that species cannot maintain its numbers, and is doomed to extinction. Thus, genetic load is assumed to place a constraint on evolution -- in particular on the maximum number of loci, whether fixed or variable, that can be maintained by selection.

Reproductive potential can be tricky to evaluate however. Viviparous organisms have physiological limits on their reproductive potential which are products of their limited resources and their large investment in the production of each offspring. Organisms that make no investment in their young beyond the production of gametes and getting them to the right place at the right time to be fertilized appear to have a much larger reproductive potential if one assumes (counterfactually) that each zygote produced can, under the best circumstances, grow to maturity. This is, of course, wildly implausible, and unapproachable except perhaps under laboratory conditions. Put this way, it seems mysterious that parental care ever evolved. This is paradoxical: parental care and other parental investment are assumed to be more "progressive" adaptations, since they increase the probability that any zygote of that type will survive to maturity. This apparently paradoxical difference is an artifact of the different stage in the life cycle at which reproductive potential is evaluated. For species without parental investment, a large amount of selection occurs between the production of zygotes and the development of young to a stage comparable to that of viviparous organisms or those which produce and care for a much smaller number of large eggs.

Including the effects of this selection yields reproductive potentials which are comparable in size.

The mechanisms we will discuss involve reduction of the genetic load by changing relevant parameters (the fourth mechanism), by converting it from one type of load to another much smaller one (the first mechanism), or by changing the way the deleterious effects act so that they do not affect the reproductive potential (the second and third mechanisms). Since each of these act to reduce the genetic load, they are not compromised by how much genetic load can be maintained, or how it should be calculated. However much genetic load can be maintained, and however it is calculated, these mechanisms would seem to be an improvement. We thus hope to avoid the controversies surrounding the theory of genetic load. We shall now discuss these mechanisms and their effects.

(1) One mechanism involves the evolutionary conversion of segregational load into mutational load. Since the mutational load per locus is generally much smaller than segregational load, many more genes can be maintained by selection if this conversion is possible. Mutational load per locus is, on different assumptions, roughly the size of the mutation rate. Segregational load per locus is roughly the magnitude of the average heterozygote advantage, and the former is normally assumed to be 2 to 4 orders of magnitude smaller than the latter (Crow and Kimura 1970).

This possibility has been suggested a number of times (Haldane 1937; Muller 1950; Crow 1952; Muller and Falk 1961; Falk 1961). A possible mechanism for this conversion involves the evolution of dominance or intermediacy from overdominance through the production of tandem duplications and recombinations which bring the contributing alleles into closely linked positions on the same chromosome and create a small "supergene." Here the segregational load is transformed into the mutational load for the new supergene, that would be roughly equal to the sum of the mutational loads for the two heterotic alleles plus the recombinational load If the genes are closely linked this could be much smaller in sum than the original segregational load. This possibility has been significantly elaborated and analyzed by Janice Spofford (1969, 1972). She notes that the large number of duplicate or closely related sequences found in metazoan genomes indicates a potentially large scope of application of this conversion mechanism.

The earlier optimism concerning the number of variable loci that can be maintained through overdominance has now waned considerably (Crow 1987), and might seem to restrict the scope and importance of this mechanism. However, the operation of Spofford's mechanism over evolutionary time could be a significant source of reduction in the number of overdominant loci. The fact that overdominant loci tend to persist for a relatively long period of time gives greater opportunity for this mechanism to act (Spofford 1969). The question remains whether it is easier to start with cooperative interactions among alleles and convert them to single loci through tandem duplication and recombination, or to find mutations which yield cooperative interactions among already existing tandem duplicates. If the former process is favored, then this mechanism should be of substantial evolutionary importance.

(2) A second mechanism involves the conversion of hard selection into soft selection (Wallace 1968) by moving the expression of crucial genes to points earlier in the developmental program, and making mutations in them early developmental lethals. This (and the next mechanism) both produce a kind of truncation selection that is often recognized as a way of reducing genetic load, (J. F. Crow pers. com.). Soft selection can occur when a bottleneck at a certain stage of the life cycle allows only a limited number of the offspring of an organism to pass, and when this number is significantly less than the number of zygotes produced. Selection occurring at earlier stages is "soft" because it does not necessarily affect the reproductive potential of the organism: it does not do so unless it removes so many organisms that less than the maximum number allowed by the bottleneck survive to that stage. Soft selection losses, as defined by the capacity at the bottleneck, do not reduce the reproductive potential of the organism.

Most organisms produce more gametes than zygotes, and more zygotes than young. Many organisms can produce only a relatively limited number of young, independent of resources, predation, and the like. Thus rats and mice conceive more embryos than they can deliver, and embryos which die early are resorbed and feed the remaining ones. Rats, in effect, apply a "second order" bottleneck by eating delivered young, thus culling the litter to the size to they can maintain and feed (M. K. McClintock pers. com.). If the lethals are expressed early enough in development, and are lost through gametic or early embryonic selection, they impose negligible reproductive costs. In effect, the genes allelic to such mutants do not participate when the fitness pie (whose size is defined by the reproductive potential) is divided among the genes that contribute to a complex adaptation expressed later in life.

Thus, a much larger number of the important genes in which mutations are lethal can be maintained through this mechanism of soft selection without significant reduction of the reproductive potential. It is important for this mechanism that the early mutations be lethal, since if they are not, the organism carrying them will pass through the reproductive bottleneck and the selective deficits they induce will be counted in the finite fitness pie divided by the forces of hard selection. Interestingly, moving the expression of a gene earlier in the developmental program enormously increases its actual or potential generative entrenchment since the scope of things that can come to depend on it is now far greater. Such a move, therefore, increases the probability that mutations in it are or will, in evolutionary time, become lethal. Thus both requirements for this mechanism to work should tend to be met together.

(3) An important variant on this second mechanism, almost certain to have occurred with appreciable frequency over extended periods of evolutionary time, is the conversion of differentially selected adaptations into "frozen accidents" through the evolution of more complex phenotypes accompanied by major increases in the generative entrenchment of these accidents. It can occur without moving the expression of such traits earlier in development, if the course of evolution towards more complex morphological phenotypes has added so much downstream of them that they *now* occur *relatively* early in development. This mechanism is meaningfully invoked only over extended periods of time, and in lineages with substantial increases in the length of development. Nonetheless, it should be important.

Frozen accidents are far more numerous than it would at first appear, since we classify as frozen accidents only those things which seem arbitrary (if adaptive), neutral, or even maladaptive (such as the vestigial vermiform appendix in man). This is in part a detection problem, since such features stand out in relief against a nicely chiseled adaptive design only if they seem obviously maladaptive. The relevant class of features includes many of what Gould and Vrba (1982) call exaptations, which they argue are far more frequent than previously supposed. A sufficiently generatively entrenched and developmentally early exaptation would be hard to detect as such because it has become such a fundamental architectural feature of a major adaptive structure that to consider its absence is tantamount to considering an alternative fundamental design.

There are many major features of adaptive design in which almost any mutations are now lethal, since so much depends upon them, but which probably were not entrenched in evolutionary periods in the distant past. Plausible candidates include the particular mappings in the genetic code; significant features of metabolism such as the Krebs cycle; mechanisms of mitosis and meiosis; mechanisms of cellular adhesion crucial to morphogenesis; and the major morphological schemata and features regarded as features of the *Bauplan*. This list is not exhaustive. If Arthur's (1984) views on megaevolution are correct, many other features should be added.

In the discussion of the first constraint, we pointed out that genes with larger selection coefficients tend to have their selection coefficients further amplified through the accumulation of dependent adaptations. We further noted that the relatively deeply generatively entrenched genes had more opportunities for the acquisition of dependent modifiers, and should thus have their selection coefficients increased more rapidly than those less significantly entrenched. These are the genes in which over time mutations should move most rapidly towards lethality. If a species has a reproductive bottleneck and some fraction of its genes move towards lethality by moving either absolutely or relatively earlier in development, (mechanisms 2 or 3), then we have an engine which uses the accumulating forces generated by the generatively entrenched genes near the top of the distribution of selection coefficients and by the possibilities provided by soft selection through mechanisms 2 and 3 to bootstrap genotypes to architectures capable of maintaining more and more genes through selection. These are architectures in which a larger fraction of the genome is relatively deeply generatively entrenched.

What about organisms which do not have a reproductive bottleneck because they do not invest any of their limited resources after producing zygotes, trusting in the weight of numbers rather than the adaptive advantages of parental care -- the large egg masses of insects or amphibia or the fertilization in sea water of the eggs of sea urchins? While they cannot derive the advantages of soft selection acting early in development, such organisms also have a much higher reproductive potential to begin with, therefore, they could tolerate larger genetic loads. To be sure, a minuscule fraction of their eggs develop to maturity, even if properly fertilized; but this is due to very intense predation, or to an unknown but presumably large number of early acting mutants -- neither of which would normally be subtracted when evaluating their reproductive potential. If such species have density dependent selection and are near their carrying capacities, most of this selection will also be "soft" (Wallace 1987), and this could provide a similar opportunity for the action of early lethals. However, it seems likely that in such organisms, a great deal of the predation is not genotype-specific. It is hard to see how, for this latter component of selection, this kind of "soft" selection could support a growth in adaptive complexity. Even so, growth in adaptive complexity in such organisms could still be driven by the first and the fourth mechanisms. This we will now discuss.

(4) One more parameter, which actually provides a fourth mechanism, is open to change. We only became aware of it when we heard Roy Britten's presentation in the Field Museum Symposium on Progress, and it is surely his story to elaborate rather than ours (Britten 1986). Britten documents significant decreases in mutation rates in some lineages over extended evolutionary time. Decreases in mutation rates are at the least important and sometimes crucial facilitating factors in allowing the evolution of more complex adaptations. As Eigen has argued, in his collaborative work on hypercycles (Eigen et al. 1981), the copying error rate for genes must be less than a threshold minimum to even support the minimal complexity necessary for the origin of a system capable of replicating genetic information.

Lower mutation rates yield lower mutational loads, in turn allow for the selective maintenance of a larger number of genes. Thus mechanisms for the reduction of mutation rates constitute another general mechanism facilitating the evolution of complex adaptations -- one which may be applicable in cases where none of the mechanisms discussed above can be significant.

Changes in mutation rate have no clear lower bounds. Conceptually, of course, mutation rates are bounded at zero, but since the halving of a mutation rate doubles the number of genes which can be maintained, this limit is without effect. It is possible, of course, that with very low mutation rates, there would be insufficient variety to drive evolution, but with any significant degree of genetic variability, recombination should be a far more significant producer of variation than mutation, so this is not a plausible limitation on the evolution of mutation rates.

The other three could allow an increase in the maximum number of genes that can be maintained by selection of 2 to 5 orders of magnitude. This depends in the first case on the ratio between average heterozygote advantage and mutation rates, and in the second and third cases on the ratio of the number of gametes of the two sexes of that type of organism (actually to the lesser of these two numbers) to the maximum reproductive potential of the organism.

These four mechanisms are "adaptation pumps" which operate over relatively long evolutionary periods and can enormously inflate the number of genes that can be maintained by selection, thus allowing the development of much more complex adaptations and of more different adaptive complexes within a given phenotype. *Thus these amount to general mechanisms for the evolution of complex adaptations.* They are general in the sense that they are not limited to any particular type of adaptation, although it remains to be demonstrated that these processes occur with significant frequency.

Summary and Conclusions

We have discussed two constraints on the evolution of complex adaptations: first, that the genes contributing to those adaptations must show variance in the magnitude of their selection coefficients, and second, that (hard) selection can maintain only a relatively limited number of such genes. We argued that the first constraint should be remarkably robust, but that the second could be circumvented in a variety of ways, over extended evolutionary time, so that during the course of evolution the number of adaptive genes maintained by selection could increase by several orders of magnitude. Such a change, we think, deserves to count as evolutionary progress.

Attempts to argue for evolutionary progress on the basis of macroscopic patterns of organic change are fraught with difficulties. On the other hand, arguments based on microevolutionary mechanisms and constraints give new handles for testing components of the hypothesis, and for evaluating the

patterns in question. Our paper is at best an interim report, and we are
working on simulations to further check, test, and elaborate these ideas.

If our sometimes speculative arguments are correct, the constraint on the
maximum number of genes maintainable by selection and the mechanisms for
avoiding it are interesting for another reason. Although they are rooted in
microevolutionary processes, these mechanisms must operate over
macroevolutionarily significant spans of time to produce a significant effect,
since they are all driven by events which are relatively rare on a
microevolutionary time scale. As such they represent mechanisms that span
two different time scales, and are the two kinds of things which are necessary
for an explanation of macroevolutionary change.

One problem with many microevolutionary mechanisms as explanations for
macroevolutionary change is that they require the maintenance of certain
conditions (e.g., isolation, overdominance, linkage, epistasis, and heritability of
fitness) whose stability may be suspect over longer periods of time. In this
case, the extrapolation from microevolutionary causes to macroevolutionary
effects is only as secure as the claims for the stability of the necessary
conditions, which may not only be suspect, but in most cases untestable. Even
if true (from a "God's eye" point of view), the status of such explanations is
likely to be indeterminate for us fallible and limited investigators. The
mechanisms proposed here are less demanding, since they utilize robust
statistical properties of the distribution of selection coefficients and their
consequences, rather than detailed properties of specific genes or sets of
genes, and their fitness histories over evolutionary time. They should
therefore be more robust, and less dependent upon the minutiae of evolutionary
history.

If correct, these mechanisms provide a basis for reintroducing a notion of
progress into evolutionary thinking -- an increasing scope of possibilities for
generating and maintaining a greater variety of complex adaptations. They do
not themselves produce complex adaptations: The production of complex
adaptations remains a product of and a subject matter for all the usual and
traditional arguments about ecological adaptation, life history strategies, etc.
Nonetheless, these mechanisms provide the possibility for the selective
production and maintenance of complex adaptations -- it seems that without
their operation, the available genetic basis could not provide the necessary

resources. Unlike nineteenth century attempts and conceptions, there is today no guarantee that man will be at or even anywhere near the apex of progressive change, but it is more likely now that there is an apex (or apices), however temporary, and that the phylogenetic tree is more than an entropic thicket, even if it is fundamentally constrained by and likely to maintain generic features of that ancestry.

Acknowledgments

James Crow, Dick Lewontin and Dave Raup commented on earlier versions, and also on a late draft of the paper. Each, Crow particularly, contributed ideas which lead to substantial changes. Most simulations were done on an Apple Macintosh and related equipment bought with earlier grants from NSF and the Systems Development Foundation. We acknowledge the support of Apple Computer for an early look at a prototype of the Macintosh II, and the National Science Foundation's History and Philosophy of Science Program, grant #SES-8709856 which gave us further access to this greater speed, and the time to design and do the appropriate simulations.

Notes

1. Fisher's fundamental theorem of natural selection, of course, provides a shorter range and more local notion of progress as increase in the mean Darwinian fitness of a population, one which is particularly well founded in population genetic theory. It fails to provide a longer range or more global notion however for several reasons: (1) Given that Darwinian fitness is a relative fitness measure, the introduction or loss of new alleles induce a rescaling of this measure, so that increase is not monotonic. (2) For similar reasons, this measure cannot be compared across populations having partially disjointed sets of alleles. For both reasons, it is not usable over extended periods of evolutionary time. (3) Since Fisher's theorem does not apply if the fitnesses are functions of gene or genotype frequency (in which case fitness will not in general be maximized by selection), and it has been widely argued

that frequency-dependent selection is very common, the theorem, at least in its original form, is of questionable applicability in real cases, and in revised form it can no longer be claimed that fitness is either increased or maximized by selection. (4) Similar remarks apply in cases with epistasis and linkage.

2. Dick Lewontin (pers. com.) has pointed out that there is some ambiguity as to how the terms "gene" and "locus" are used by geneticists. To a population geneticist, a gene is an allele -- one of a variety of sequences (usually of DNA bases) which can occur at a given locus or place in the chromosome, and which can be moved around, deleted, duplicated, etc. A locus is a site in a chromosome which can be occupied by various alleles, and which, in polyploid organisms has homologous sites occupied by the same or different (usually similar) alleles. To many other geneticists, genes and loci are conceived not in terms of chromosomal location but primarily in terms of function -- the classical *cistron*. A locus or a gene is that which produces a given protein product which is all (or part of, for enzymes which contain a number of similar subunits) an enzyme which catalyzes a given biochemical reaction, or a control gene which turns on a gene or genes at other loci.

In this paper, genes and loci are used in both senses in a (hopefully) consistent manner, on the assumption that the genes we are interested in could be characterized in either fashion. (This rules out, for example "junk" or silent DNA that can be characterized according to the first criterion, but not to the second.) Discussions of genetic load obviously refer to the first sense, but talk of evolutionary processes which reduce the genetic load per locus by converting hard selection into soft selection, or of genes which acquire dependent adaptations are obviously presupposing claims about functions of genes, changes in these functions, or changes in the time of expression of these functions. The first mechanism (for converting segregational load into mutational load through tandem duplication and recombination) seems to use primarily the first sense of gene, and the fourth (reduction of mutation rate) could be characterized in either the first or the second sense, depending on the mechanism for the reduction of mutations.

3. James Crow has suggested a significant reformulation of this idea. His preferred formulation of this measure of progress is "... *an increase in*

selectable genetic variability per unit genetic load." Crow observes that this
revision makes it more capable of precise specification [measurement], and
experimental test (pers. com.). His proposal or one like it would be required,
not only for experimental or observational tests, but also for analogous tests
of the idea through computer simulation. This proposal has the effect,
however, of changing the emphasis of the argument: We seek to explain the
large number of fixed loci that can be maintained by selection processes, not
the number of loci that contribute significant genetic variation. The latter is,
of course, the concern of the classical discussions of genetic load, heterosis,
and the maintenance of genetic variability. (See Lewontin 1974, and Crow's
1987 and Wallace's 1987 historical reviews of these debates.) If the load at
these fixed loci can be reduced, i.e., through conversion to mutational and/or
recombinational load for closely linked loci, then this also allows for the
maintenance of a greater number of segregating loci, so his proposal and ours
are not unconnected. Since our mechanisms work through the fixation of
segregating alleles, together with the selective and developmental consequences
of this fixation, we must, at some stages consider the case of variable loci.
However, we do not, for these mechanisms, require the maintenance of
substantial genetic variability through selection, so the classical arguments
over heterosis and load have no place here. See below.

4. Lewontin (1978) also advances a related requirement, which he calls the
"principle of continuity" -- that small phenotypic changes do not generally lead
to major changes in fitness, in which case even small quasi-independent
changes in phenotype could lead to strong interactional effects in (and
consequent low heritability of) fitness. As Lewontin points out (pers. com.), if
the principle of quasi-independence can be thought of as a genetic or inward-
looking requirement, the principle of continuity can be thought of as an
environmental, ecological, or outward-looking requirement. They can also be
thought of two sides of the requirement that variance in fitness be additive or
heritable, a perspective that fits well with Crow's (1986, pp. 194-95) discussion.

5. Not only must they produce benefits without scrambling a significant part
of the adaptive organization of the phenotype, resulting in at least a tolerable
degree of fitness, but unless their fitness in heterozygous form is greater than

that of the alternatives, these new types must strong positive assortative mating, occur in small isolates, or in sufficiently high frequency together with sterility or greatly reduced fitness when mated with the extant types, or some sufficient combination of these, to become established. Arthur (1984) explicitly makes some or all of these assumptions.

6. This rough equilibrium is technically a "moving equilibrium" in the sense of Lotka (1924), in which some processes have a faster relaxation time than others, and therefore come to an equilibrium for given values of the parameters of the other processes. However this second set of parameters is also changing, albeit more slowly, so that the equilibrium value now changes at the slower rate determined by the slower processes. This phenomenon is common in hierarchically organized systems in which higher level processes change more slowly than lower level ones.

7. This list is not intended to be exhaustive, even among general conditions or mechanisms, and we have specifically not included mechanisms which act to increase the amount of genetic variability, such as high mutation and recombination rates or stabilizing frequency-dependent selection.

8. This differs slightly from the standard population genetic definition of genetic load in terms of the ratio of the difference between maximum and realized population fitness to the maximum fitness (Crow and Kimura 1970). We have chosen to define genetic load in terms of reproductive potential (an absolute fitness measure) to render the arguments more direct. Both definitions will produce the same numerical values, and nothing in our arguments depends on this difference.

References

Arnold, A. J., and K. Fristrup. 1982. The theory of evolution by natural selection: A hierarchical expansion. *Paleobiology* 8:113-29.

Arthur, W. 1984. Mechanisms of morphological evolution: A combined genetic, developmental and ecological approach. New York and London: John Wiley and Sons.

Brandon, R., and R. Burian, eds. 1984. *Genes, organisms and populations.* Cambridge: MIT Press.

Britten, R. J. 1986. Rates of DNA sequence evolution differ between taxonomic groups. *Science* 231:1393-98.

Burian, R. 1986. On integrating the study of evolution and of development. In *Integrating scientific disciplines,* ed. W. Bechtel, 209-28. Dordrecht: Martinus-Nijhoff.

Crow, J. F. 1952. Dominance and overdominance. In *Heterosis,* ed. J. Gowen, 282-297., Ames, Iowa: Iowa State College Press.

Crow, J. F. 1986. *Basic concepts in population, quantitative, and evolutionary genetics.* New York: W. H. Freeman.

Crow, J. F. 1987. Muller, Dobzhansky, and overdominance. *Journal for the History of Biology* 20, No. 3 (Fall):351-80.

Crow, J. F., and M. Kimura. 1970. *An introduction to theoretical population genetics.* New York: Harper.

Darwin, C. [1859] 1965. *The origin of species.* Facsimile reprint of 1st edition. Cambridge: Harvard University Press.

Eigen, M., W. Gardiner, P. Schuster, and R. Winkler-Oswatitsch. 1981. The origin of genetic information. *Scientific American* 244 (April):88-118.

Falk, R. 1961. Are induced mutations in *Drosophila* overdominant? II. Experimental results. *Genetics* 46:737-57.

Gould, S. J. 1982. The meaning of punctuated equilibrium and its role in validating a hierarchical approach to macroevolution. In *Perspectives on Evolution,* ed. R. Milkman, 83-104. Sunderland, MA: Sinauer.

Gould, S. J., and R. C. Lewontin. 1978. The spandrels of San Marco and the panglossian paradigm: A critique of the adaptationist programme. *Proceedings of the Royal Society of London* 205:581-98.

Gould, S. J., and E. Vrba. 1982. Exaptation --A missing term in the science of form. *Paleobiology* 8:4-15.

Haldane, J. B. S. 1937. The effect of variation on fitness. *American Naturalist* 71:337-49.

Kauffman, S. A. 1969. Metabolic stability and epigenesis in randomly constructed genetic nets. *Journal of Theoretical Biology* 22:437-67.

Kauffman, S. A. 1985. Self-organization, selective adaptation and its limits: A new pattern of inference in evolution and development. In *Evolution at a crossroads: The new biology and the new philosophy of science,* ed. D. J. Depew and B. H. Weber. Cambridge: MIT Press.

Kauffman, S. A. 1986. A framework to think about evolving genetic regulatory systems. In *Integrating scientific disciplines,* ed. W. Bechtel, 165-84. Dordrecht: Martinus-Nijhoff.

Lewontin, R. C. 1970. The units of selection. *Annual Review of Ecology and Systematics* 1:1-18.

Lewontin, R. C. 1974. *The genetic basis of evolutionary change.* New York: Columbia University Press.

Lewontin, R. C. 1978. Adaptation. *Scientific American* 239(3)(September):212-30.

Lotka, A. J. [1924] 1957. *Elements of mathematical biology.* Reprint. New York: Dover Books.

Muller, H. J. 1950. Our load of mutations. *American Journal of Human Genetics* 2:111-76.

Muller, H. J., and R. Falk. 1961. Are induced mutations in Drosophila overdominant? Experimental design. *Genetics* 46:727-35.

Olby, R. [1966] 1985. *The origins of Mendelism.* 2d ed. Chicago: The University of Chicago Press.

Rasmussen, N. 1987. A new model of developmental constraints as applied to the Drosophila system. *Journal for Theoretical Biology* 127 3(August 7):271-301.

Rosenblatt, F. 1962. *Principles of neurodynamics.* Washington: Spartan Books.

Schank, J. C., and W. C. Wimsatt. In press. Generative entrenchment and evolution. In *PSA-1986, vol. 2,* ed. P. K. Machamer and A. Fine. Lansing, MI: The Philosophy of Science Association.

Spofford, J. 1969. Heterosis and the evolution of duplications. *American Naturalist* 103:407-32.

Spofford, J. 1972. A heterotic model for the evolution of duplications. In *Evolution of genetic systems,* ed. H. H. Smith, 121-143. Brookhaven Symposia in Biology, no. 23. New York: Grodon and Breach.

Wade, M. J. 1976. Group selection among laboratory populations of Tribolium. *Proceedings of the National Academy of Sciences* 73:4604-07.

Wade, M. J. 1977. An experimental study of group selection. *Evolution* 31:134-53.

Wade, M. J. 1978. A critical review of the models of group selection. *The Quarterly Review of Biology* 53:101-14.

Wallace, B. 1968. *Genetic load.* Englewood Cliffs, NJ: Prentice-Hall.

Wallace, B. 1987. Fifty years of genetic load. *The Journal of Heredity* 78:134-42.

Williams, G. C. 1966. *Adaptation and natural selection: A critique of some current evolutionary thought.* Princeton: Princeton University Press.

Wilson, D. S. 1980. A theory of group selection. *Proceedings of the National Academy of Sciences* 72:143-46.

Wilson, D. S. 1980. *The natural selection of populations and of communities.* Menlo Park, CA: Benjamin/Cummings.

Wilson, D. S., and E. Sober. 1987. The superorganism concept. Unpublished.

Wimsatt, W. C. 1980. Reductionistic research strategies and their biases in the units of selection controversy. In *Scientific discovery, Vol. II: Case Studies,* ed. T. Nickles, 213-59. Dordrecht: Reidel.

Wimsatt, W. C. 1981. Units of selection and the structure of the multi-level genome. In *PSA-1980, vol. 2,* ed. P. D. Asquith and R. N. Giere, 122-83. Lansing, MI: The Philosophy of Science Association.

Wimsatt, W. C. 1986. Developmental constraints, generative entrenchment, and the innate-acquired distinction. In *Integrating scientific disciplines,* ed. W. Bechtel, 185-208. Dordrecht: Martinus-Nijhoff.

Wright, S. 1931. Evolution in Mendelian populations. *Genetics* 16:97-159. 1986. Reprinted in *Sewall Wright -- evolution: Selected papers,* ed. W. B. Provine, 98-160. Chicago: The University of Chicago Press.

Evolution, Progress, and Entropy

E. O. Wiley

The problem of determining if evolution is progressive lies with the problem of defining progress itself. How can evolution be progressive without a goal to progress toward? Can the "goal" be anything but an observer-dependent goal? We might do better to ask ourselves a different kind of question. What can we discover about evolution that can be accounted for by more general, perhaps physical, laws? Brooks and Wiley (1986, and earlier papers cited therein: Collier 1986, 1987; Wiley 1987; Brooks, Collier, and Wiley 1986) suggest that evolution is characterized by increasing entropy accompanied by increasing complexity and increasing organization. Evolution is yet another manifestation of the operation of the second law of thermodynamics. However, the entropy functions (or, more correctly, the partial entropy functions) associated with organismic diversification are not the usual entropy functions encountered in the thermodynamic behavior of nonliving systems. Rather, they are partial entropy functions associated with the genetic code and other relevant aspects of the information systems of organisms. In other words, biological information (in all of its hierarchical manifestations) is subject to the constraints of the law of entropy increase in the same way that energy flows are subject to the law in its thermodynamic manifestation. This conjures up the possibility of certain paradoxes. For example, on an intuitive level we tend to associate greater complexity or greater organization with negentropy (i.e., a decrease in the entropy of a system that is necessary to achieve the nonrandomness that we term organization). However, these apparent paradoxes disappear when we discover that the behavior of biological information systems corresponds to a particular subset of dynamic models that allow information to be dissipated over lineages at a rate faster than it can be bound.

Energy and Information

Everyone agrees, I presume, that organisms are far-from-equilibrium dissipative structures. Organisms require a continuous input of free energy in order to maintain themselves in a steady state. Organisms "obey" the energy/matter aspects of the second law in a manner that is not fundamentally different from that of nonliving systems of a similar (if far simpler) kind. For example, steam engines and organisms function only because they are provided with (or acquire) a continuous supply of free energy. However, this very fact undermines any thought that the strictly thermodynamic aspects of the law of entropy increase should lend insight into the processes involved in biological diversification. In other words, if organisms "obey" the thermodynamic aspects of the second law just like steam engines, then strictly thermodynamic considerations are not likely to tell us much about why there are millions of species of organisms but only one basic kind of steam engine.

This conclusion is reinforced when we consider "information." Organisms differ from nonliving "systems" in that organisms contain what Wiley and Brooks (1987) term "instructional information." As Collier (1986, 1987) has characterized it, instructional information is comprised of a physical array of information whose properties depend only on the properties internal to the system in question. It is no less real than Brillouin's (1962) "bound information" (similar to the "structural information" of Wiley and Brooks 1987). At the DNA level, instructional information is represented by a variety of levels of organization from triplet codes, to "genes," to genotypes. The properties of this instructional information depend on the properties intrinsic to DNA molecules, not on the surroundings or environment. To put it another way, the environment does not "mold" DNA molecules. While all physical structures can be thought of as having structural information that is a manifestation of their structural complexity, only organisms have instructional information. As a loose analogy, organisms carry their blue prints inside and constantly refer to them, while the blue print of a steam engine stays on the engineer's desk.

Organisms build themselves by referencing their own intrinsic, instructional information. This information comes from one or more parents in the form of DNA, chromosomal organization, and cytoplasmic organization (that is directly traceable to translation and transcription in a parent). The structures we observe can be directly related to these instructions. ("Directly related" should not be construed as a statement claiming a one-to-one correspondence between phenotype and genotype, only that there is a definite and observable link between genes and the similarities and differences between organisms. The statement is meant to be trivial.)

The Brooks and Wiley (1986) theory differs from other so-called thermodynamic theories in claiming that information takes precedence over energy when we consider the impact of the law of entropy increase on organisms. We can see this point when we consider the fact that it is the instructional information which determines the kinds of chemical reactions that will occur at a rate fast enough to maintain the steady state of an organism in a particular environment. In other words, instructional information is not only directly related to structural organization, it also determines how energy will flow through the organism. Brooks and Wiley (1986) do not assert that energy flow is trivial, only that there is no thermodynamic imperative behind organismic diversification. Thus, our view of the relationship between the strictly thermodynamic aspects of the second law of thermodynamics and evolution are not much different from those (like Morowitz 1968) who claim that since organisms are open thermodynamic systems, there are no particular energy constraints on increasing biological complexity and organization. Rather, we suggest a new approach: instructional information is subject to the constraints of the second law in a similar manner to the constraints placed on energy flow and this is of fundamental importance to understanding the basic processes of evolution.

Information

Instructional information exists as a physical array (Collier 1986). It can be duplicated, transcribed, translated, and transmitted from generation to generation through reproduction. Instructional information fits neither the concept of information proposed by Shannon and Weaver (1949) nor the concept proposed by Brillouin (1962; see Collier 1986). Instructional information is not observer dependent (i.e., phenomenological), contrary to most concepts of "information" encountered in the physical sciences. In other words, instructional information exists in the absence of an observer. The physical reality of instructional information should come as no surprise to biologists because the reality of both molecular and Mendelian genetics depends upon biological (instructional) information being a real physical array. The physical reality of biological information may come as a distinct surprise to information theorists and/or physical scientists, who tend to treat "information" as synonymous with "description." The distinction between biological information as a physical array and phenomenological information as description is, in a fundamental sense, the difference between a property of a system (that is subject to physical constraints) and a mere account of a system (that is observer dependent and thus of no particular relevance to any natural law).

The physical aspects of instructional information in organisms are quite different from the physical aspects of matter/energy exchange. Organisms are open systems in terms of matter/energy exchange, but they are closed systems in terms of information exchange. In other words, while organisms require a continuous exchange of matter/energy with the environment to maintain a steady state, they do not require a continuous exchange of instructional information with the environment to maintain a steady state. Indeed, the environment cannot provide instructional information to an organism.

Given that instructional information resides in organisms, comprises a physical array, and is "closed," we may ask: what are the consequences of the dynamics of translating, transcribing, duplicating, and transmitting such an array relative to a general physical law that predicts that entropy will increase during any real series of processes?

Entropy and Information

Brooks and Wiley (1986, and earlier papers) suggest that a number of partial entropy functions are associated with various levels of biological organization. Further, they have asserted that these entropy functions are concerned with the information systems of organisms and not with matter/energy exchange (which is dependent on the information systems). The validity of these assertions depends on (1) recognizing the physical reality of instructional information (contrary to Wicken 1987), (2) our ability to distinguish between informational microstates and macrostates (a necessary distinction if one is to apply the formalisms of statistical mechanics to physical information systems), and (3) the technical requirement that the macrostate itself does not determine the microstates that define the phase space of organisms.

A microstate is a state which might be occupied by a particular entity (cf. a particle or an organism). It is a class to which an entity or a number of entities comprising a system might belong. A macrostate is the distribution of the entities of a system over the microstates available to the system. The statistical entropy of a physical system is directly related to the macrostate and thus to the configurational complexity of the system. The arena of possible configurations is directly related to the number of available microstates. This is termed the phase space, and is defined as

$$H_{max} = \log_2 A$$

where A is the number of available microstates.

Layzer (1977, 1978, 1980) has suggested that the phase space of biology is a genetic phase space. This phase space is real in the sense that instructional information is real (i.e., that genes, genotypes, and phenotypes are real). We may distinguish between microstates and macrostates because biological systems are hierarchical. The instructional information at one level provides the microstates for the next higher level in the hierarchy. As Collier (submitted) has stated: "Assuming that the solution [of microstate/macrostate distinctions] exists for the chemical level (Holzmuller 1984), the problem is to give an account of the entropy associated with information at a given level higher in the hierarchy of information."

Consider a very simple case, a one locus-two allele system. The phase space may be defined as the number of possible genotypes. Each genotype class is a microstate. The population (N organisms) comprises the system. The macrostate is the distribution of these individuals over the microstates available to the population. The entropy is a measure of this distribution and can be measured directly by the usual Shannon formula (a modification of the

Boltzmann formula using $\log_2$ rather than $\log_e$, rendering $k = 1$).

$$H = -\sum(p_i \log_2 p_i)$$

where p_i is the probability of a particular organism occupying a particular genotype. The result is a measure of the configurational complexity of the population in terms of the distribution of individuals over the microstates available to a particular system at a particular hierarchical level of information (in this case, the level represented by genotypes). It is a measure of the entropy of information. For simple Mendelian systems, Ginsberg (1981) has shown that this measure (which he terms the entropy distance) has macroscopic properties.

At this point I assume that I have demonstrated that two of the three requirements have been met: (1) Instructional information comprises a physical array that can be interpreted under the second law. (2) Microstate/macrostate distinction can be made. Requirement (3) is easily met in biological systems. A population cannot conjure up the microstates available to it. It is the immediate past generation that determines the microstates available to the population plus the number of copy mistakes that have occurred in the gametes. (Copy mistakes are, or course, purely entropic phenomena in the flow of information since they represent a randomization process relative to the previous state of the system and increase the phase space of the system if they are novel.)

Neither Brooks and Wiley (1986), nor Collier (1986, 1987) argues that the phase space, microstates, or macrostates described in our papers are classical thermodynamic microstates. The entropies are array entropies, more like the entropies of sorting encountered in considering an ideal gas than to the thermal entropies associated with steam engines. This does not mean that the entropies we propose are unreal, mere metaphors for "real" entropies. Far from it. It means that what we term the second law of thermodynamics is more general than the narrow interpretation usually allowed.

Brooks and Wiley (1986) suggest that the behavior of genetic systems undergoing change shows positive increases in the partial entropy functions associated with them. In other words, we suggest that evolution is an entropic phenomenon, a natural consequence of the behavior of physical

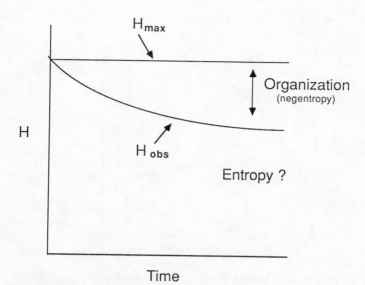

Figure 1. The relationship between a constant entropy maximum (H_{max}) and the observed entropy of a system (H_{obs}) over time for a physical information system. The difference is a measure of organization and has been interpreted as negentropy. H_{max} is measured as $\log_2 A$, where A is the number of microstates available to the system. H_{obs} is measured by the Shannon formula $H_{obs} = -\sum(p_i \log_2 p_i)$.

information arrays in an entropic universe. This may seem paradoxical. Common sense teaches us that evolution is characterized by (1) increasing complexity, (2) increasing organization, and (3) increasing order, giving the illusions of progress and of negentropic behavior.

The key to solving this apparent paradox lies in the realization that evolution is a nonequilibrium phenomenon characterized by an increasing phase space and a tendency for realized variation to lag behind an entropy maximum (i.e., for realized diversity to lag behind maximum possible diversity). Consider two ways of viewing the unfolding of evolution (figs. 1 and 2). In the first (fig. 1), the observer is perched on a line (labeled $\log_2 A$) that represents an entropy maximum. The observer sees, at any one time, that realized genetic diversity is lower than maximum possible diversity. Further, over time, the observer sees that realized diversity seems to recede away from the observer. The observer concludes that evolution is a negentropic phenomenon since

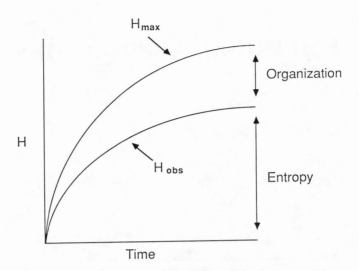

Figure 2. The relationship between an increasing entropy maximum ($H_{max} = \log_2 A$) and the observed entropy $H_{obs} = -\sum(p_i \log_2 p_i)$ of a physical information system over time. The difference is organization while the value of H_{obs} is a measure of the entropy of the system.

greater organization requires greater departures from randomness. (Greater organization is measured as the difference between randomness, the $\log_2 A$ line, and observed complexity, the H_{obs} line; see Brooks and Wiley 1986.)

In the second view (fig. 2), the observer is also perched on the line that represents entropy maximum. The observer sees, as time goes by, that realized genetic/organismic diversity increasingly lags behind the ever-increasing entropy maximum. Indeed, the gap between H_{max} and H_{obs} at any one time is the same as that in the previous view. However, in this case the observer is also moving, because the entropy maximum is increasing over time. Further, the entropy maximum is increasing at a faster rate than the realized diversity. The observer, realizing this, concludes that evolution is characterized by entropic behavior, and also realizes that increases in complexity and increases in organization are, themselves, emergent properties of this entropic behavior.

The models shown in figures 1 and 2 are phenomenologically identical. In both cases the observer sees the realization of diversity receding from her favored place on the curve of entropy maximum. However, figure 1 is biologically unrealistic. This model postulates a constant entropy maximum.

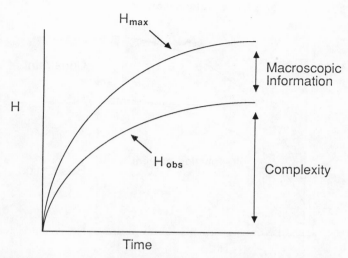

Figure 3. The relationship between macroscopic information and complexity of a physical information system under the Brooks/Wiley theory.

For us to accept a constant entropy maximum, we must postulate that no new genetic variation has been generated since the origin of life. We must postulate that the diversity we observe is precipitated out of a constant and unchanging array of primeval variation.

The second model (fig. 2) is biologically realistic. It postulates that there has been a global increase in genetic diversity over time. This does not mean that every lineage and every species must show such increases. Rather, it means that such increases are hypothesized to occur over the sum of all lineages. This model conforms to a class of thermodynamic models discussed by such workers as Layzer (1977), Frautschi (1982) and Landsberg (1984a, b). In such models the order of a system can increase as the number of accessible microstates increases so long as the realized entropy increases at a slower rate than the maximum possible entropy (graphically, the result in fig. 2). Using this model, we can define some relevant terms.

The difference between the entropy maximum (H_{max}) and the observed entropy (H_{obs}) is a measure of the organization of the system (fig. 2). Following Layzer's (1977) terminology, this difference is the macroscopic information (fig. 3). It is also a measure of constraint (fig. 4). A Shannon

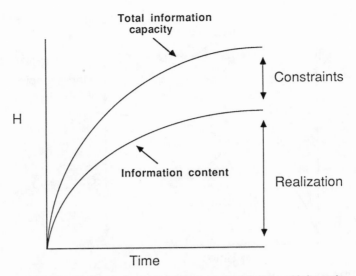

Figure 4. The relationship between total information capacity (H_{max}) and information content (H_{obs}) in a physical information system. The difference between total information capacity and information content provides a measure of the constraints placed on the information system. The entropy of information (H_{obs}) gives a measure of the amount of realized information diversity.

measure (H_{obs}, figs. 2-4) is a measure of the entropy (fig. 2) of the physical information system (or an observable manifestation of that system). It is also a measure of complexity (fig. 3), or realized diversity (fig. 4) of the system. In terms of constraints, the distance between H_{max} and H_{obs} represents possible variation that has been historically excluded (fig. 5), while H_{obs} stands for variation that has been historically realized. In terms of organization, the difference between H_{max} and H_{obs} represents a measure of the bound information (Collier 1986).

Some Biological Interpretations

The partial entropy functions proposed by Brooks and Wiley (1986) can be applied to a variety of levels of biological organization where one or more manifestations of instructional information are functional (i.e., manifested in the form of heritable attributes or transcribed products). While most of these

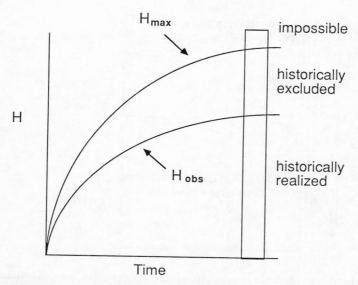

Figure 5. The relationship between H_{max} (total information capacity) and H_{obs} (information content) of an array of physical information systems comprising a number of evolutionary lineages. Historically realized diversity is measured by H_{obs}. Historical exclusion of the expression of certain kinds of information is measured by $H_{max} - H_{obs}$. The area above H_{max} represents an area of impossible combinations because no novelties exist (at the time observed). Note that what is impossible at one time period may become possible at a later time period. Also note, however, that evolution is characterized by an ever increasing area of historically excluded combinations.

levels directly relate to biological evolution, we would assert that the general model applies to all biological phenomena that show irreversible behavior involving components of instructional information. As Collier (submitted) stresses, these are partial entropy functions. However, this presents no problems since partial entropy functions are additive across and between hierarchical levels. Four examples, based on the discussions of Wiley and Brooks (1986) are listed below:

1. Ontogeny. Consider a developing multicellular organism. During its development the total number of cell types increases as a result of cellular differentiation. We can measure this by taking the $\log_2$ of the total number of cell types at any particular time. These values, represented by the upper curve in figure 6, signal increases in the phase space of cell types and define an upper limit on the cellular complexity of the organism at any particular

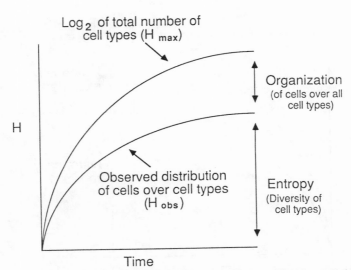

Figure 6. The relationship, during growth and development of a organism, between the total number of cell types (Log $_2$ of the total, or H_{max}) and the observed distribution of cells over cell types (H_{obs}). The organization represents the ontogenetic constraints and history of the pattern of development. The entropy represents the complexity of the organism.

time. The number of cells also increases, but at a faster rate than cellular differentiation, producing a nonrandom distribution of cells over cell types. We can measure this value with the standard Shannon formula. If there is no organization to the growing organism, we would expect the Shannon value to equal the entropy maximum. That is, the organism would be riding the upper line. However, we do not observe this; rather, we see that the observed distribution of cells is nonrandom over the cell types and that there are more cells than cell types. The configurational entropy of the organism and its complexity both increase, as shown by the lower curve. The constraints placed on differentiation and the distribution of cells over cell types also increase, and these are manifested in the increasing organization of the organism. This is shown by the increasing lag between the upper and lower curves. Please note there is no reason to think that the strictly thermodynamic entropy of the organism would do anything but increase along with the informational entropy since increasing structural complexity requires increasing energy to maintain the steady state. However, it is possible that the rate of increase per unit (such as per cell) might drop -- a manifestation of the organization of the whole (see Brooks and Wiley 1986).

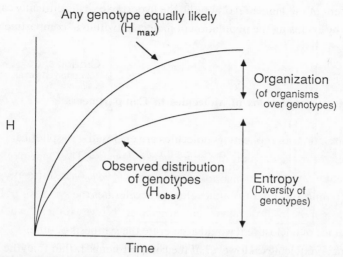

Figure 7. The relationship, among an array of organisms, between the diversity of possible genotypes and the observed distribution of organisms over genotypes. If the array is riding the upper line (H_{max}), then any genotype is as likely as any other. If the observed distribution of genotypes is nonrandom, as in the lower (H_{obs}) line, then the array is organized to an extent measured by the difference between the two lines. The H_{obs} curve is a measure of the diversity of genotypes.

2. Genotypic evolution. Over phylogenetic space (i.e., the entire tree of life) we observe an increasing number of possible genotypes as evolutionary novelties occur. If all of these possible genotypes were equally likely to be encountered in nature, then we would expect evolution to be at an entropy maximum (the upper curve in fig. 7). Instead, we observe a lag between the number of possible genotypes and the distribution of observed genotypes (shown by the increasing distance between the upper and lower curves in fig. 7). One reason for this increasing lag lies in the fact that many genotypic combinations (and phenotypic expressions) are rendered impossible due to the fact that speciation acts to partition genetic information between lineages (fig. 8). Another reason is that history, manifested in the bound information of lineages (Collier 1986; Brooks and Wiley 1986) and expressed in the ontogenetic constraints acting during the development of individuals, constrains the successful realization of genotypes even though they might represent accessible microstates. This represents natural selection in its internal, developmental

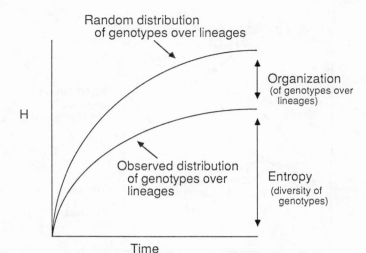

Figure 8. One reason that genotypes are nonrandomly distributed over arrays of organisms is that the organisms themselves are distributed over independently evolving lineages. Thus, there will be a difference between the observed distribution of genotypes over lineages (lower curve) and the theoretical expected distribution of genotypes over lineages under the assumption that any genotype is as likely in any lineage.

mode. Another reason, is course, is that natural selection in its environmental mode eliminates certain phenotypes and their underlying genotypes.

3. Character evolution. We can see the gross aspects of this phenomenon when we consider character evolution. Consider three species whose characters are shown in the table in figure 9. The characters that evolved during the common ancestry history of these species (i.e., during the evolution of these species and their immediate common ancestors) are shown as "+" signs while characters that evolved before the origin of their common ancestor are shown as "-" signs. If there were no phylogenetic constraints on character evolution, we might expect the evolutionary history of the characters and taxa to be like that shown on the tree to the lower right. If there are phylogenetic constraints, then a history like that on the lower left is more reasonable. The difference is graphically shown in the upper plot. Phylogenetic analysis (Hennig 1966; Wiley 1981) under the Brooks/Wiley theory is a protocol that strives to estimate the true configurational entropy of a given clade for the characters examined (see Brooks, O'Grady and Wiley 1986, for a measure of this property). Derived characters represent bound

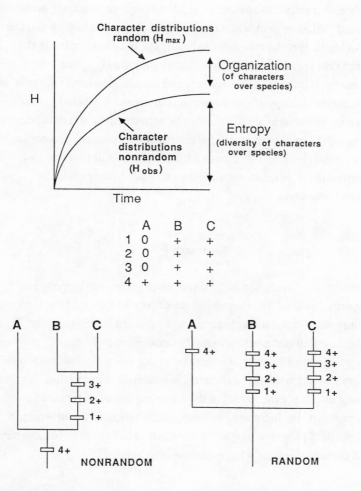

Figure 9. Systematic implications of the Brooks/Wiley theory. *Upper*, a graph showing the expected relationship between random and actual character distributions over lineages in a clade. *Middle*, a character matrix for three species comprising a strictly monophyletic group where "+" represents an apomorphy and "-" represents a plesiomorphy. (Character argumentation is assumed to be true.) *Bottom left*, the distribution of apomorphies over the relationships of taxa as predicted by the Brooks/Wiley theory and implemented by using Hennigian argumentation. A value of the entropy of information for this tree would be predicted to fall on the H_{obs} line of the upper graph. *Bottom right*, the distribution of apomorphies under the assumption that there are no historical constraints on the evolution of characters, i.e., that all similar characters might be independently derived. A value of the entropy of information for this tree would fall on the H_{max} (upper) line in the upper graph.

information and the extent to which this bound information also provides
ontogenetic constraints is the extent to which subsequent character evolution
is constrained. What we seek when estimating phylogenetic history and the
evolution of characters is a true estimate of H_{obs}. If we overestimate the
entropy of the clade (i.e., our measure is above the true H_{obs} line), we
overestimate the number of convergences and parallelisms in our data. If we
underestimate the entropy of our clade (i.e., our measure is below the true
H_{obs}), then we underestimate the number of convergences and parallelisms.
Either error will have consequences for our study of evolution. Interestingly,
I know of no previous theory that gives a theoretical account for what we
should be striving for when we estimate evolutionary history using
phylogenetic techniques.

Summary

The direction of evolution is the direction of increasing entropy and
increasing organization. The thermodynamic arrow of time and the biological
arrow of time point in the same direction and suggest a common law of
entropy increase that is shared between thermodynamic systems and physical
information systems. However, the increase in organization that is observed in
evolutionary systems is not due to increasing tendency to perfection, nor does
it signal progress. It is a byproduct of the historical constraints placed on
species by past history. Increasing entropy is antithetical to the nineteenth
century concept of progress. Evolution may look teleological or orthogenetic
because of these constraints -- but evolution is neither.

Acknowledgments

Thanks are due to Daniel R. Brooks and John Collier for discussions of
physical arrays, to David Hull for discussions of the general aspects of the
Brooks and Wiley theory, and special thanks to Matthew H. Nitecki and the
Field Museum of Natural History for inviting me to participate in the Spring
Systematics Symposium.

References

Brillouin, L. 1962. *Science and information theory*. New York: Academic Press.

Brooks, D. R., and E. O. Wiley. 1986. *Evolution as entropy. Towards a unified theory of biology*. Chicago: University of Chicago Press.

Brooks, D. R., J. D. Collier, and E. O. Wiley. 1986. Definitions of terms and the essence of theories: A reply to J. S. Wicken. *Syst. Zool.* 35(4):640-47.

Brooks, D. R., R. T. O'Grady, and E. O. Wiley. 1986. A measure of the information content of phylogenetic trees, and its use as an optimality criterion. *Syst. Zool.* 35(4):571-81.

Collier, J. D. 1986. Entropy in evolution. *Biol. Philos.* 1:5-24.

Collier, J. D. 1987. The dynamics of biological order. In *Information, entropy and evolution*, ed. D. Depew, B. Weber, and J. Smith. Cambridge: MIT Press.

Collier, J. D. Submitted. Defining entropy: A reply to Wicken. *Philos. Sci.*

Frautschi, S. 1982. Entropy in an expanding universe. *Science* 217:593-99.

Ginsberg, L. R. 1981. *Theory of natural selection and population growth*. Menlo Park, CA: The Benjamin/Cummings Publ. Co.

Hennig, W. 1966. *Phylogenetic systematics*. Urbana: University of Illinois Press.

Holzmuller, W. 1984. *Information in biological systems: The role of macromolecules*. Cambridge: Cambridge University Press.

Landsberg, P. T. 1984a. Is equilibrium always an entropy maximum? *J. Stat. Physics* 35:159-69.

Landsberg, P. T. 1984b. Can entropy and "order" increase together? *Physics Lttrs*. 102A:219-20.

Layzer, D. 1977. Information in cosmology, physics and biology. *Int. J. Quantum Chem.* 12 (Suppl. 1):185-95.

Layzer, D. 1978. A macroscopic approach to population genetics. *J. Theor. Biol.* 73:769-88.

Layzer, D. 1980. Genetic variation and progressive evolution. *Amer. Natur.* 115:809-26.

Morowitz, H. J. 1968. *Energy flow in biology: Biological organization as a problem in thermal physics*. New York: Academic Press.

Shannon, C. E., and W. Weaver. 1949. *The mathematical theory of communication*. Urbana: University of Illinois Press.

Wicken, J. S. In press. Entropy and information: Suggestions for common language. *Philos. Sci.*

Wiley, E. O. 1981. *Phylogenetics: The theory and practice of phylogenetic systematics*. New York: Wiley-Interscience.

Wiley, E. O. 1987. Entropy and evolution. In *Information, entropy and evolution*, ed. D. Depew, B. Weber, and J. Smith. Cambridge: MIT Press.

Wiley, E. O., and D. R. Brooks. 1987. A response to Professor Morowitz. *Biol. Philo.* 2:369-74.

Testing the Fossil Record for Evolutionary Progress

David M. Raup

It is reasonable to ask whether the fossil record shows evidence of successively "better" organisms through geological time. And the lavish record of some 250,000 fossil species provides an ample sample with which to test ideas of evolutionary progress. The problem should be reducible to two simple questions: (1) What should be measured? and (2) How should the data be analyzed? The situation is far more complex, however, and it is unclear whether the presence or absence of evolutionary progress can be demonstrated rigorously. This conclusion is somewhat discouraging but it constitutes a challenge for future researchers to develop more effective approaches to the problem.

In this paper, I will review some of the problems and pitfalls inherent in using the historical record to study progress. Then, I will consider a single case in some detail: the observation that taxonomic groups show increasing resistance to extinction through the Phanerozoic.

What Can be Measured?

Table 1 lists several measurable quantities often or usually contained in paleontological data and which, intuitively at least, might be appropriate as metrics of evolutionary progress. Not included are the many aspects of organismic biology which might show progress but which are inaccessible to the paleontologist. For example, one might predict evolutionary progress in the effectiveness of the immune system but data on immune systems of extinct species are totally lacking.

Table 1

Examples of possible metrics of progress measurable in the fossil record.

MORPHOLOGIC

> Single characters (body size, surface/volume ratio)
> Complexity (information content, number of parts)
> Variability (single characters or groups of characters)
> Adaptive success (optimization of structural systems)
> Adaptive innovations (evolution of flight, mimicry, intelligence)

CLADOGENETIC

> Taxonomic diversity (species richness, species/family ratio)
> Branching rate (frequency of speciation, origin of new higher taxa or
>   body plans)
> Survivorship (change in species or genus half-life)

ECOLOGIC

> Community structure (change in complexity or stability of communities)
> Exploitation of adaptive zones (marine reef development, occupation of
>   terrestrial environments)
> Biogeography (change in geographic range or provinciality)
> Coevolution (development of angiosperm-insect relationships)

Some the items in Table 1, such as community structure, are elusive in fossil data because of sampling limitations or lack of preservation of whole biotas. Others provide abundant data but analysis is difficult. An example is biogeography: even though most fossil species are reasonably well located in space and time, changes in physical geography over geologic time (continental drift) make analysis difficult.

By far the most attractive metrics of progress in the fossil record are the simplest ones such as body size and duration of lineages. It is thus not surprising that evolutionary increase in body size (Stanley 1973; LaBarbera 1986) and changes in taxonomic survivorship (Raup and Sepkoski 1982) have been used most often to investigate evolutionary progress.

Problems of Time Series Analysis

To search for evolutionary progress is inevitably to work with time series. The question often asked is: Do the data (temporal changes in some metric) show a significant trend over geologic time?

There are two fundamentally different kinds of fluctuating time series in nature, one based on a series of independent events and the other based on some sort of markov process. In the first case, values of a metric are drawn independently from some probability distribution and can be exemplified by the sequence of throws of dice, each count having a constant probability of being thrown so that the time series fluctuates up and down and shows no persistent trends.

In the markov case, on the other hand, the position of the metric at a given time is constrained by its previous position. An example is the changing fortune of a gambler who makes a series of relatively small wagers. In any time interval, the gambler's fortune may increase or decrease but is constrained to be relatively close to that existing at the start of the interval. That is, if a gambler has $100 and bets $5 in a game with a 1:1 payoff, the outcome can only be to increase the stake to $105 or reduce it to $95.

Virtually all imaginable evolutionary time series are of the markov type in that the state of the system at any time is equal to the state at the immediately preceding time plus or minus some relatively small quantity. The significance of this in our search for evolutionary progress is that *most markovian time series show trends and these trends are often monotonic (nonreversing).* Thus, constancy in historical data is the exception. Furthermore, it is often the case that increases or decreases over time are deemed to be "statistically significant" in the sense that if the record were sampled again, the same increase or decrease would probably emerge. But because directional change is the a priori expectation for almost any markovian time series, documenting change does not necessarily tell us anything about "progress."

Formal statistical analysis of time series is outside the expertise of most biologists, and there are some special problems, as shown by Yule (1926) and by Feller (1950) in classic treatments of the subject. Consider, for example, the simple random walk shown in figure 1. The "walk" starts at time = 0 (on the left) and moves through 1000 time steps (to the right). At each of these steps, the walk has a 50-50 chance of going up or down and the resulting track is shown in the plot. This particular random walk is typical in that it appears to be "going somewhere." In fact, random walks which give an appearance of aimless wandering are surprisingly rare. If we naively apply

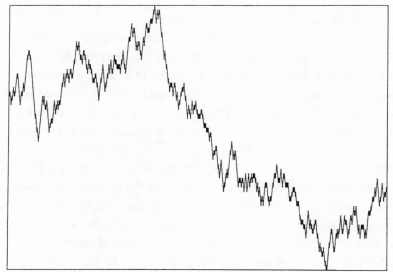

Figure 1. Computer-generated random walk of 1000 steps. At each step, the walk has a
50-50 chance of going up or down by a unit distance. There appears to be a "statistically
significant" trend with time (downward, in this case) but this is typical of random walks.

standard statistical techniques to this random walk, we find that the position
of the walk (vertical dimension in the plot) is significantly correlated with
time. That is, a correlation coefficient (r) for the 1000 pairs of data is
significantly different from zero (see Feller 1950, for discussion and more
examples).

The prevalence of apparently significant directional change is an
unfortunate characteristic of random walks and this influences our intuitive
interpretation of real time series as well as their formal statistical analysis.

It is important to emphasize that the comparison of real time series to
purely artificial random walks is not to imply or claim that nature is random.
Rather, the experience with random walks provides a vital null expectation --
the pattern one expects to find if *nothing of consequence* is driving the
system. This possibility must be ruled out before we can begin to interpret
observed trends in the context of progress.

Figure 1 shows historical change in only one dimension but the same
phenomenon is common in higher dimensions as well. Figure 2 is reproduced
from an earlier study of random evolution in the morphology of imaginary

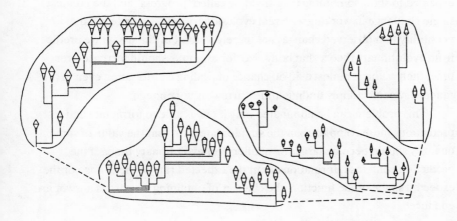

Figure 2. Simulated phylogeny of imaginary "triloboids." Each triloboid in the evolving system is drifting randomly in six dimensions and the simulation thus represents a 6-dimensional random walk. Note that there are clear trends in shape and size even though there is no directional pressure in the computer algorithm. Modified from Raup 1977, Fig. 5.

organisms called triloboids (Raup and Gould 1974). At each branch point in a simulated phylogeny, each of six linear dimensions is subject to constant probabilities of a unit increase, a unit decrease, or no change. Thus, triloboid form can stay the same or change in shape and/or size, size being the sum of the dimensions. As can be seen in figure 2, a number of persistent changes in shape and size do take place throughout the phylogeny and this, in the real world of the fossil record, could be interpreted as directional change in response to persistent driving forces, forces known to be absent in this example.

The triloboid example is, in effect, a six-dimensional random walk. An important difference between this and the one-dimensional walk is that as dimensions are added, the probability of a return to a previously occupied state rapidly approaches zero. That is, in random walks of more than a few dimensions, the historical record rarely repeats itself. Therefore, the null expectation of directional change becomes even stronger as more complex structures are considered.

Thus, any time series, regardless of the number of dimensions, can be expected to show something that might be called "progress" and the greatest single challenge in working with real evolutionary data is to rule out the possibility that observed changes are merely a result of chance. If we predict from evolutionary theory that body size, for example, should increase over time, then we have almost a 50-50 chance of finding confirmation even in a random system. Simply finding "time's arrow" is not enough.

This problem can be eliminated only if one is able to formulate and test predictions more complex than the simple statement that the value of some quantity should become larger or smaller over evolutionary time. If the prediction contains information about the expected rate of the change or the expected shape of the function, the chances of obtaining a rigorous answer are enhanced.

Pull of the Recent

The fossil record is distorted by a complex of time-dependent biases collectively called the "Pull of the Recent" (Raup 1979). As one goes back in time, the number of fossils goes down as the area and volume of sedimentary rock available for study decreases. Numbers of exposures of sedimentary rock decline because earlier rocks have a higher probability of having been eroded or covered. Also, because of the greater likelihood of diagenetic or metamorphic alteration of fossiliferous sediments, the earlier rocks generally yield less well preserved fossils and considerable morphological information is lost. There are exceptions to these generalizations, of course, but the statistical trends through time are obvious and important.

In the present context, the Pull of the Recent is especially serious because it can induce trends we might be tempted to call progress. For example, the increase toward the Recent in both quantity and quality of the fossil record makes an increase in *apparent* species diversity almost inevitable *even in the absence of any real increase in number of species*. This makes analysis of changes in diversity (species richness) extremely difficult -- although a reasonable consensus has developed in support of the idea that diversity has indeed increased during the Phanerozoic (Sepkoski et al. 1981).

The Pull of the Recent has subtle but significant effects on our assessment of changing morphology through time. Because the rocks deposited later in geologic time generally have more fossils, they also contain more of the morphological variability originally present. The null expectation, therefore, is that coefficients of variation will increase toward the Recent and this can easily appear as a successful test of an evolutionary prediction.

Even apparently simple traits can be influenced by the bias of geologic time. Many traits, including body size, show strongly right-skewed frequency distributions so that parameters such as the *mean* and the *maximum* increase with sample size. If sample sizes increase over time because of improvement in preservation or rock exposure, then the null expectation may be that observed means and maxima (but not medians) will also increase. Some of these problems may be accommodated through careful attention to sample size and quality of preservation.

An insidious but important element of the Pull of the Recent is the preference among taxonomists to assign fossil species to groups of living organisms. Such assignments, when incorrect, make fossil groups appear to have survived much longer than they actually have, and this has the effect of greatly exaggerating taxonomic diversity in the later parts of the record. This effect is especially crucial in any analysis of changing patterns of survivorship and will be discussed more fully later in this paper.

Kinds of Predictions to be Tested

The literature of evolutionary biology and paleobiology is replete with explicit or implicit suggestions about the kinds of progress that should be expected under the Darwinian paradigm of evolution. For example, it is normal and conventional to assume that organisms should become more diverse -- morphologically, ecologically, and taxonomically -- through time. They should become better adapted (optimized) and become better survivors, leading to an increase in species durations. Communities should become larger, or more stable or more complex. And the organisms themselves should become more complex.

Some or all of these predictions are suspect, to say the least, and their roots in solid evolutionary theory are often difficult to find. Some of the predictions are surely more the product of an attempt to explain the history of evolution as it exists than the result of solid theory. Consider, for example, the geologic record of species diversity: the changing standing crop of species as measured through the proxy of changing numbers of genera and families. All published analyses agree that taxonomic diversity has increased by several orders of magnitude since the early Precambrian and that the increase over the past 600 million years (Phanerozoic time) has been substantial -- by at least a factor of three or four (Sepkoski et al. 1981). Furthermore, it is widely agreed that the fossil record shows the increase continuing to the present, with present diversity of the total biota being higher than at any time in the history of life.

The observed increase in taxonomic diversity seems totally credible as a product of evolution. After all, we know evolution to be a branching process wherein a species may give rise to many other species, so the basic mechanism for diversification is available. In addition, we know that the evolutionary invasion of new environments (such as the land or the air) has opened up space for diversification. And coevolution, such as that between insects and flowering plants, has made the potential for further diversification almost limitless. It is thus straightforward to make a prediction of increasing diversity.

But how might our thinking be different if the actual geologic record showed a *decrease* in taxonomic diversity from, say, the Carboniferous onward? I suspect that this could be accommodated by theory and might even be seen as the most logical prediction from theory. One could argue, for example, that there should be an early evolutionary stage of experimentation in organismic design accompanied by (or made possible by) high diversity but that this should be followed by a reduction in diversity caused by the elimination, by extinction, of all but the most nearly optimal or hardy species. In fact, precisely this framework has been used to explain why taxonomic diversity within individual clades shows a strong statistical tendency to be high early in their history, as shown by Gould et al. (1977) and more recently by Gilinsky and Bambach (1986). Is there any reason why a prediction from theory which is found to work with single biological groups should not apply also to the whole biota?

My point is not to criticize specific predictions that have been put forward but to emphasize the risk of inadvertently deriving the theoretical predictions of evolutionary change from the very historical record we seek to explain.

One measure of the power of our theoretical framework to make testable predictions about evolutionary progress is to ask: If the historical record were inverted, would we recognize it as being in conflict with theory? For example, there are several well-established trends in the evolution of trilobites in the Paleozoic. Trilobites appear in the early Cambrian record with a remarkably high morphologic and taxonomic diversity. For the Cambrian as a whole, roughly three-quarters of all fossil species known are trilobites. Most of the largest and most complex trilobites (as well as many of the smallest and simplest) occur in the first third of the Paleozoic. As we move through the Paleozoic, however, numbers of species dwindle and the range of morphologic diversity decreases until finally, in the Permian, the group becomes extinct.

I do not mean to suggest by the trilobite example that our fossil record may, in fact, be upside down, because we know from biostratigraphy and from radiometric dating that it is in the correct time sequence. And good phylogenetic studies should be capable of establishing direction even in the absence of a geologic time scale. But in terms of the sorts of metrics we might use to evaluate evolutionary progress, we must be wary of the possibility that claims of progress may be made even with an inverted chronology. This emphasizes the point that the predictions from theory must be precise, fairly complex, and include a clear temporal polarity.

Problems of Scale

If one views the history of life from a great distance and looks at just a few points, there is obvious directionality. Compare, for example, the biology of the Precambrian of 2.0 billion years before present (B.P.) with that of the Recent. At 2.0 B.Y.B.P. there were few if any eukaryotic organisms and diversity was extremely low by any measure. Many environments and adaptive zones were unoccupied. The contrast with the modern biota is dramatic and few observers would deny that progress of some sort has taken place.

As we work with shorter time spans and use more closely spaced sample

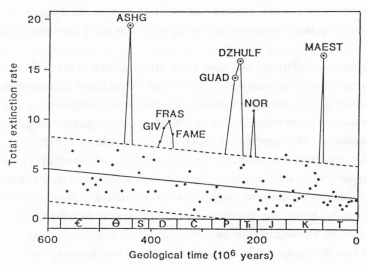

Figure 3. Decline in the rate of background extinction for families of marine animals. Regression line is fit to those points having extinction rates <8 families per myr, thus eliminating the several mass extinctions that stand above background extinction. Dashed lines define the 95% confidence band for the regression. Modified from Raup and Sepkoski 1982, Fig. 1. Abbreviations refer to conventional geologic time units.

points, however, recognizable directionality becomes elusive. At some point in this change of scale, interpretable change is either completely absent or lost in noise caused either by sampling effects or by epi-phenomenal reactions to short-term changes in conditions. It is important, therefore, to choose the scale of an investigation of evolutionary progress with great care.

Evolutionary Progress through Enhanced Survivorship

In a paper designed primarily to locate mass extinctions in the fossil record of marine families, Raup and Sepkoski (1982) noted an irregular but persistent decline in background extinction rate from about 5.0 families/million years in the Cambrian to about 2.0 families/million years in the Tertiary (fig. 3). The observed decline was striking because the increase in standing diversity through the Phanerozoic should have been expected to yield an *increase* in extinctions per million years. We suggested that the decline may be evidence of evolutionary progress through improved resistance to extinction.

Several paleontologists have commented on the study and have offered alternative interpretations. Flessa and Jablonski (1985) noted several indications that speciosity of families has, on the average, increased during the Phanerozoic and suggested that the observed reduction in family extinction rate could be an artifact of increased numbers of species per family. Indeed, various compilations of diversity records show that the rate of increase in species diversity through the Phanerozoic has exceeded that for family diversity (Valentine 1969) so that the total number of species per family (as a statistical average) must have increased also. More detailed data are difficult or impossible to find, however, because of severe limitations of the species fossil record.

More recently, Boyajian (1986) showed that the age structure of the assemblage of families changes systematically and predictably through the Phanerozoic. Cambrian families cannot be very old because most of them originated in the Cambrian itself. But as one moves through time, the assemblage of families can and does "age" because some fraction of the first-formed families have survived. Boyajian argued that because older families have, on the average, more species, they are more resistant to extinction and that this provides an explanation for the observed decline in family extinction rate. The aging effect is significant on Phanerozoic time scales because family durations constitute a significant fraction of the total duration of the Phanerozoic.

The Flessa-Jablonski and Boyajian studies cast doubt on our original interpretation of "progress" and this has inspired the more extensive analysis to be presented here. Although the results remain equivocal, the case is a valuable one to illustrate the serious problems inherent in any search for progress in the fossil record of past life.

Durations of Fossil Species

It is recognized that some species survive longer in geological time than others. The variation is large and highly skewed, with the greatest number of species having durations substantially less than the mean. Furthermore, it has been shown in several cases that survivorship curves for fossil species are approximately log-linear, indicating a constant probability of extinction throughout the life of the species (Van Valen 1973).

Probably most of the variation in species duration observed within a genus or family is purely stochastic and is analogous to survivorship at the population level in those organisms that show no age-dependent mortality (passerine birds, for example). That is, if species always have the same risk of extinction, actual durations vary widely as a matter of chance. But as Van Valen (1973) and others have shown, some biologic groups have consistently higher or lower extinction rates than others.

Species duration is clearly an appropriate metric of evolutionary progress because it is a direct and explicit measure of the success of a species in the struggle to avoid the ultimate failure: extinction. Furthermore, Jablonski (1987) has shown that those characteristics of organisms that promote long-term survival are often strongly heritable. And the longer a species survives, the more likely it is to produce progeny in the form of other, similar species.

It should be noted in passing that the species extinction discussed here is *true extinction*, the termination of a species lineage, rather than what is called *pseudoextinction*, wherein a species is transformed by phyletic evolution into a descendent species. As will be shown, the separation of the two kinds of extinction is not critical because any general analysis must be carried out at higher taxonomic levels where the problem of pseudoextinction is rarely a problem.

Survivorship at Higher Taxonomic Levels

Unfortunately, *species* durations are rarely measurable in the fossil record because of limitations of preservation and time resolution. The cases reported by Van Valen (1973) involved a few groups of microfossils where the data were reasonably complete, but these are quite exceptional. We must, therefore, attempt to use durations of higher taxa as proxies for species-level data. Sepkoski's (1982) compendium of the geologic ranges of families has greatly facilitated such studies.

If one is willing to make certain assumptions about the stochastic nature of the cladogenetic process, it is possible to compute species extinction rates (and therefore mean or median durations of species) from survivorship data on higher taxa. Formulas for this conversion have been applied to fossil genera by Raup (1978, 1985).

It is much more straightforward, however, simply to use survivorship of higher taxa as a substitute for species survivorship. This has some problems and potential pitfalls. First, durations of higher taxa change nonlinearly with species durations: a doubling of species duration, for example, will more than double the duration of the higher taxon to which the species belong. This problem is minor and can be accommodated. More fundamental problems involve (1) the questionable validity of the higher taxonomic groupings, and (2) the effects of variation in the number of species per higher taxon (the effect noted by Flessa and Jablonski 1985).

On the first problem, it is often argued that the higher taxa conventionally used in paleontology are defective because they are either polyphyletic or paraphyletic. Whereas it can be granted that phylogenetically incorrect (polyphyletic) groupings are common in taxonomy, it is also true that for many kinds of fossil organisms, the products of conventional taxonomy serve to define valid groupings in that the contained species *are* closely related. This is seen most strikingly in instances where a completely new set of traits (often nonmorphological) is surveyed in an established group and variation is found to conform to existing taxonomy. As long as one is working with statistical patterns in large samples, the vagaries caused by polyphyletic groupings are probably minimal.

The question of paraphyly is more serious. Many, perhaps most, higher taxa established by traditional methods of evolutionary taxonomy are paraphyletic in the sense that the groups do not include *all* the descendants of the group. A family may appear to become extinct but actually survive through descendent species that are defined as originators of new genera or families. This phenomenon is, in a loose sense at least, the equivalent of species pseudoextinction (above) but at higher cladogenetic levels. Nevertheless, the traditional families still serve well in defining groups of related species occupying single adaptive zones or small portions of morphospace. To cite a classic example, the traditional taxonomic groupings used for dinosaurs serve the useful purpose of delineating a distinctive kind of animal that died out at the end of the Cretaceous. This is true even though such groupings are paraphyletic because the descendants of one or more members actually survive today as birds. It seems quite legitimate to ask, therefore, whether survivorship of such groups, paraphyletic or not, has "improved" over geologic time.

The second problem noted above is the effect of change in the number of species per higher taxon. All other things being equal, a family or genus will persist longer if it contains more species merely because more species extinctions are required to effect its demise. Thus, an increase in speciosity of genera (or "generosity" of families) through the Phanerozoic could produce an increase in the mean durations of the higher taxa. Such an increase might well be the result of an increase in speciation rate rather than any increase in resistance to extinction. This problem is indeed severe and must be accounted for in any analysis of the survivorship of higher taxa.

Extinction at the Generic Level

The problems encountered with families can be greatly reduced by working at the level of the genus. There are approximately ten times more genera than families defined in the Phanerozoic record and this sample size makes possible kinds of analysis not possible at the family level. Also, mean durations of genera are a much smaller fraction of total Phanerozoic time so that the aging effect noted by Boyajian is of lesser significance.

Sepkoski is in the process of compiling a large dataset of generic ranges and this now numbers about 30,000 taxa. The July, 1986, version of this dataset will be used in the analyses that follow.

Cohort Survivorship Curves for Genera

As shown by Van Valen (1973) and subsequent workers, survivorship curves for species and higher taxa often provide a sensitive measure of the varying tempo of extinction. Figure 4a shows a nested set of cohort survivorship curves for a subset of 15,582 genera of the Sepkoski dataset.

Each descending curve monitors the decay of a cohort by the extinction of its genera. The cohorts are "true" cohorts in that they consist of all the genera originating in one of the generally recognized Phanerozoic stages. This is in contrast to the "pseudocohorts" or "polycohorts" used where sample sizes are limited (Hoffman and Kitchell 1984; Raup 1986). Each of the survivorship curves in figure 4a provides an independent means of tracking extinction behavior over a segment of the Phanerozoic.

Before plotting figure 4a, genera with certain kinds of time ranges had to be eliminated in order to avoid the biasing effects of the Pull of the Recent. In brief, all Cenozoic cohorts were ignored as were all "extant" genera of the older cohorts. This can be justified with reference to Table 2.

Table 2 shows the numbers of extinctions by stage in several of the Sepkoski cohorts. For each of the stages from the Aptian (Cretaceous) onward, the numbers of genera becoming extinct in each cohort are given (columns) as well as the number of genera alleged to be living today (bottom row labeled "EXTANTS"). Extinctions in a given cohort decline to near zero within a few stages, indicating that the cohorts are virtually exhausted in that time. However, many of the cohorts have a substantial number of "extant" genera. This is almost certainly a result of the Pull of the Recent due to invalid assignment of fossil species to living groups yet it is impossible to quantify the effect rigorously.

The approach of removing all "extants" and all post-Cretaceous cohorts should eliminate the problem, however. This means being willing to assume that true generic survival for more than 65 million years is unlikely. The half life for the average Phanerozoic genus, not counting the effect of the late Permian mass extinction is slightly more than 10 million years (Raup 1978). Still, the average rate of species turnover for Phanerozoic is such that about 10% of the genera in a cohort should remain after 65 million years. That this is not the case is due to the presence of the terminal Cretaceous mass extinction which reduced cohort sizes to a point where they are exhausted during the subsequent 65 million years, and this is the main rationale for filtering out all genera alleged to be extant and all post-Cretaceous cohorts. The filtering means a considerable loss of information, but it is justifiable as the only effective way of removing the strong biasing effect of the Pull of the Recent.

Table 2

Numbers of extinctions in 20 generic cohorts of post-Barremian stages. For pre-Tertiary cohorts, numbers of extinctions dwindle almost to zero as the Recent is approached. This indicates that the cohorts are almost completely extinct after 65 million years yet the generic range data indicate substantial numbers of members still alive ("EXTANT"). For pre-Tertiary cohorts, most of the "EXTANTS" are artifacts of the "Pull of the Recent."

Mesozoic cohorts   Cenozoic cohorts

STAGE COHORT SIZE	Apt 204	Alb 313	Cen 371	Tur 173	Con 127	San 134	Cam 285	Mae 291	:	Dan 128	Tha 146	E-1 201	E-m 424	E-u 294	O-1 141	O-u 121	M-1 291	M-m 202	M-u 159	Pli 359	Ple 166
EXTINCTIONS:									:												
Aptian	68	-	-	-	-	-	-	-	:	-											
Albian	28	147	-	-	-	-	-	-	:	-	-	-	-	-	-	-	-	-	-	-	-
Cenomanian	25	61	178	-	-	-	-	-	:	-											
Turonian	6	13	35	66	-	-	-	-	:	-	-	-	-	-	-	-	-	-	-	-	-
Coniacian	1	3	3	13	37	-	-	-	:	-											-
Santonian	5	1	5	7	20	48	-	-	:	-	-	-	-	-	-	-	-	-	-	-	-
Campanian	3	9	17	5	13	23	75	-	:	-	-	-	-	-	-	-	-	-	-	-	-
Maestrich.	31	31	60	42	29	34	132	182	:	-	-	-	-	-	-	-	-	-	-	-	-
Danian	0	4	14	2	2	1	15	17	:	44	-	-	-	-	-	-	-	-	-	-	-
Thanatian	1	2	1	1	1	2	3	11	:	8	33	-	-	-	-	-	-	-	-	-	-
Lower Eoc.	1	0	2	1	0	0	2	3	:	2	6	41	-	-	-	-	-	-	-	-	-
Mid Eocene	0	3	3	0	1	2	1	3	:	2	2	14	129	-	-	-	-	-	-	-	-
Upper Eoc.	1	4	1	1	1	0	3	4	:	2	14	18	53	89	-	-	-	-	-	-	-
Lower Olig.	0	0	2	0	1	0	1	0	:	2	5	2	9	9	38	-	-	-	-	-	-
Upper Olig.	0	1	1	1	0	0	0	1	:	1	1	1	4	6	1	28	-	-	-	-	-
Lower Mio.	0	1	1	1	0	0	1	0	:	1	2	3	10	6	8	8	62	-	-	-	-
Mid Miocene	0	0	0	0	2	0	0	2	:	2	3	5	13	12	7	6	32	81	-	-	-
Upper Mio.	2	0	0	0	0	1	1	0	:	0	2	3	2	7	3	3	5	9	69	-	-
Pliocene	0	1	2	1	1	0	1	1	:	4	7	7	5	5	3	2	16	9	13	104	-
Pleist.	1	0	0	0	0	0	0	0	:	0	0	1	0	0	0	0	3	2	6	14	15
"EXTANT"	31	32	47	33	19	22	50	67		60	71	106	199	160	81	74	173	101	71	241	151

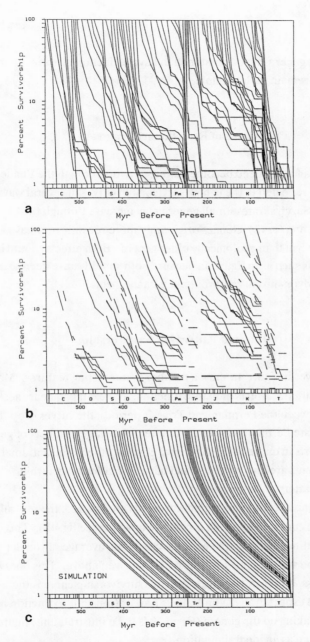

Figure 4. Cohort survivorship curves for Phanerozoic genera. Only pre-Tertiary cohorts are used to avoid the bias of the Pull of the Recent. (*a*) Survivorship curves for 15,582 fossil genera from data provided by J. J. Sepkoski. (*b*) Same plot with the steep portions of curves eliminated so that the effects of mass extinction are eliminated. (*c*) Simulation of ideal cohort survivorship of genera using Phanerozoic averages for species turnover rates.

Also, the lost genera can be presumed to have extinction histories comparable to those that remain in the sample.

Predictions of Evolutionary Progress

If genera have indeed become better survivors through the Phanerozoic, this might be reflected in the cohort survivorship curves in several ways, as follows: (1) Survival rates for all genera may increase through the Phanerozoic, and/or (2) Genera originating more recently may have a better record of survival than geologically older genera. Both effects should be visible in plots such as in figure 4a because both should cause decreasing slopes of survivorship curves through the Phanerozoic.

Observed Patterns of Survivorship

The curves in figure 4a show a number of interesting features. Most pronounced are the effects of mass extinction. For example, near the boundary between the Permian and Triassic periods, the curves drop sharply because the late Permian mass extinction removes large numbers of genera from all cohorts in existence at that time. Several other large and small extinction events are also recorded and have a significant perturbing effect on larger scale patterns of survivorship.

If the "scarps" formed by extinction events are ignored, the overall shape of most of the curves in figure 4a is concave upward (see fig. 4b). This means that extinction rates for genera decrease, on average, through the life of a cohort and suggests one of the predictions given above. This does *not* prove progress, however, because a concave survivorship curve is the mathematical expectation if Van Valen's Law of Constant Extinction holds for the species making up the genera (Raup 1978). To illustrate this, figure 4c was constructed as an ideal simulation.

In figure 4c, imaginary cohorts were constructed for each stage in the Phanerozoic time scale assuming a constant rate of generic origination. Each genus in each newly formed cohort was assumed to have a single founding species and the subsequent fate of the genus depended on the balance of speciation and species extinction within that genus. Probabilities of species origination and extinction were taken from an earlier study (Raup 1978) and were held fixed at 0.081 and 0.09, respectively.

The simulation was done analytically as a fully deterministic process using the following relationship between species turnover and generic survival (from Raup 1978, 1985):

$$\text{Proportion of genera surviving} = 1 - \frac{q(e^{(p-q)t} - 1)}{p\,e^{(p-q)t} - q}$$

where p is the rate of origination of congeneric species, q is the rate of species extinction, and t is the time (in millions of years) since the founding of the cohort. The form of the equation requires that the survivorship curve is concave-upward because the longer a cohort survives, the lower the probability of extinction of its members in the next time interval.

The pattern of survivorship in figure 4c is thus what we might see in an ideal world free of episodic extinction and free of irregularities due to sampling error. The only irregularities are the occasional bunchings of lines caused by irregularities in stage duration in the geologic time scale. The pattern in figure 4c can be used as a null hypothesis against which the real world of figure 4a can be compared. If the real record shows evolutionary progress in survivorship, we should expect to find a decrease in average curve slope *and/or* a decrease in the initial slope from left to right in figure 4a.

It should be noted in passing that the null model (fig. 4c) seems to show a *steepening* of the slopes of survivorship curves from left to right but this is merely an optical illusion caused by the truncation of curves on the right.

Does Overall Extinction Rate Decrease Through the Phanerozoic?

Figure 5a is a plot of average percent generic extinctions per stage using

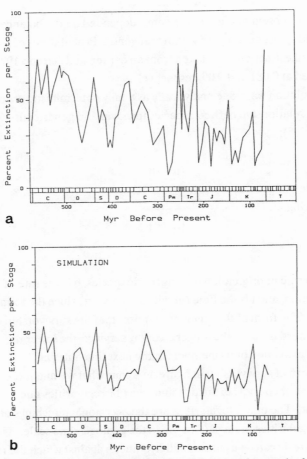

Figure 5. (*a*) Percent extinction per stratigraphic stage based on the records of 15,582 fossil genera from data provided by J. J. Sepkoski. (*b*) Simulation of the same plot using a model wherein speciation rates and species extinction are held constant. Irregular fluctuations in the simulated curve are due to variation in stage duration. The long-term decline in generic extinction is due primarily to the reduction in average stage duration through the Phanerozoic.

all 15,582 generic ranges contributing to figure 4a. Although the trend is ragged because of perturbing extinction events, there is clearly a decline in rate through time. But this is strongly influenced by the fact that stage duration in the conventional geologic time scale also decreases. In the null model, shorter stages inevitably have fewer extinctions. A comparable plot using data from the simulation (fig. 5b) shows a similar decline, with

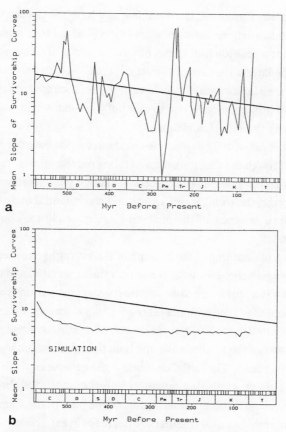

Figure 6. (*a*) Weighted mean slope of survivorship curves in fig. 4a with fitted regression line. Fluctuations in mean slope are due to mass extinctions in some stages and to inaccuracies in geologic dating (stage durations) in others. (*b*) Simulated mean survivorship curve slope from data in fig. 4c compared with the regression line from fig. 6a.

irregularities being due to the vagaries of variable stage duration.

The problem of stage duration can be eliminated by normalizing for time. This is done in figure 6a: a plot of percent generic extinction per million years for the real data. Because all genera present in each stage were considered, this plot tracks, in effect, the weighted average of the slopes of all the survivorship curves in the original graph. Figure 6a shows extremely irregular fluctuations. The upward spikes are caused by mass extinctions, most noticeable in the late Cambrian, late Permian, and terminal Cretaceous, *or* by stage durations that are unrealistically short in the Harland time scale used here. The spike at a position four stages before the end of the Cretaceous is an example of the latter: the Coniacian stage is credited with only one million years in the Harland scale. Downward spikes, such as that for the first stage of the Permian, may be due to unusually low extinction rates or to unrealistically long time stage duration.

In spite of the noise in figure 6a, one *could* infer a downward trend in extinction rate through the Phanerozoic and this is represented by the regression line that has been superimposed. Although the fluctuations about the line are extreme, the trend, *if real*, constitutes a substantial increase in survivorship from a generic half-life of about 7 myr on the left to about 15 myr on the right.

Is the apparent doubling of the half-life of the average genus real? This can be tested by again comparing the trend with the expectations of the null model. Figure 6b is a plot of the same metric (percent extinctions per million years) for the simulated data with the regression line from figure 6a plotted above. The simulated data show a rapid decrease in extinction rate for the first 100 or 150 myr of the Phanerozoic and then no significant change for the remainder of the record. The initial decrease is the generic equivalent of Boyajian's aging effect: extinction rate must be higher in the Cambrian because no genera can be older than the time to the base of the Cambrian.

It is indeed problematical whether the decrease in extinction rate observed in figure 6a is a real biological phenomenon or merely the trend produced by the null model (fig. 6b) with the addition of severe noise. Statistical tests could probably be developed to support either alternative. The interpretation of the observed decline must, therefore, remain equivocal.

Survivorship of Cohorts in Their First Stage

One of the predictions of progress made earlier was that genera originating more recently should have a better record of survival than genera originating earlier. To test this, the slopes of survivorship curves for cohorts in their *first stage* are tracked in figure 7. Again, there is considerable noise

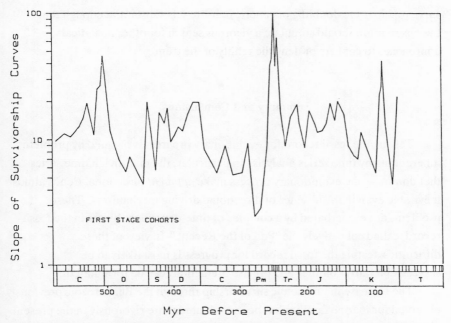

Figure 7. Changing slope of survivorship curves for cohorts in their *first* stage. Some of the irregularities are due to changes in extinction rate and some are due to inaccuracies in geologic dating.

because of mass extinctions and errors in the time scale (stage durations). But even though the time series is extremely irregular, figure 7 gives no visual indication of a downward trend in extinction rate.

Is There Progress in Generic Survivorship?

We have seen several indications of substantial improvement in survivorship (decrease in extinction rate) during the Phanerozoic but in each case, valid arguments can be made that the trends are not significant or that they are artifactual. Only the decline in mean slope of survivorship curves (fig. 6a) remains a reasonable candidate and this is questionable. In spite of the noise seen in figure 6a and in spite of the possibility that it is not separable from the null expectation (fig. 6b), the net improvement in survivorship -- the doubling of generic half-life -- is substantial enough that the observation should stimulate a vigorous search for other analytical approaches to confirm or deny the reality of the trend.

Summary and Conclusions

Testing the fossil record for evolutionary progress is plagued by problems inherent in any time series analysis. In particular, the markovian time series that dominate the evolutionary process make *apparent* directional trends almost inevitable even in the absence of directional driving mechanisms. These problems are exacerbated by a complex of time-dependent biases in the fossil record, called collectively the "Pull of the Recent." In view of these difficulties, testing the fossil record for progress is most likely to be successful if fairly complex and precise predictions can be formulated.

The increase in taxonomic survivorship through the Phanerozoic previously observed for marine families has been evaluated more rigorously in the present analysis using Sepkoski's new dataset for genera. The generic data also show a substantial increase in survivorship, expressed by a doubling of generic half-lives from the Cambrian to the Recent. However, this may be an artifact of taxon aging, coupled with extremely noisy data, as indicated by the fact that the survivorship of generic cohorts in their first stage of existence does not increase through time.

With taxonomic survivorship, as with other possible metrics of evolutionary progress, unequivocal answers may have to wait for the development of a more precise geologic time scale and more effective means of eliminating the Pull of the Recent.

Acknowledgments

Some of the research described here was supported by Grant NAG-237 of the National Aeronautics and Space Administration (USA). I thank J. W. Valentine and an anonymous reviewer for helpful criticisms of the manuscript.

References

Boyajian, G. F. 1986. Phanerozoic trends in background extinction: Consequence of an aging fauna. *Geology*, 14:955-58.

Feller, W. 1950. *An introduction to probability theory and its applications*. Vol. 1. New York: John Wiley and Sons.

Flessa, K. W., and D. Jablonski. 1985. Declining Phanerozoic background extinction rates: Effect of taxonomic structure? *Nature* 313:216-18.

Gilinsky, N. L., and R. K. Bambach. 1986. The evolutionary bootstrap: A new approach to the study of taxonomic diversity. *Paleobiology* 12:251-68.

Gould, S. J., D. M. Raup, J. J. Sepkoski, Jr., T. J. M. Schopf, and D. S. Simberloff. 1977. The shape of evolution: A comparison of real and random clades. *Paleobiology* 3:23-40.

Hoffman, A., and J. A. Kitchell. 1984. Evolution in a pelagic, planktic system: A paleobiologic test of models of multispecies evolution. *Paleobiology* 10:9-33.

Jablonski, D. 1987. Heritability at the species level: Analysis of geographic ranges of Cretaceous mollusks. *Science* 238:360-63.

LaBarbera, M. 1986. The evolution and ecology of body size. In *Patterns and processes in the history of life*, ed. D. M. Raup and D. Jablonski, 69-98. Berlin: Springer-Verlag.

Raup, D. M. 1977. Stochastic models in evolutionary palaeontology, In *Patterns of evolution*, ed. A. Hallam, 59-78. Amsterdam: Elsevier Sci. Publ. Co.

Raup, D. M. 1978. Cohort analysis of generic survivorship. *Paleobiology* 4:1-15.

Raup, D. M. 1979. Biases in the fossil record of species and genera. *Carnegie Museum of Natural History Bulletin* 13:85-92.

Raup, D. M. 1985. Mathematical models of cladogenesis. *Paleobiology* 11:42-52.

Raup, D. M. 1986. Biological extinction in earth history. *Science* 231:1528-33.

Raup, D. M., and S. J. Gould. 1974. Stochastic simulation and evolution of morphology -- Towards a nomothetic paleontology. *Systematic Zoology* 23:305-32.

Raup, D. M., and J. J. Sepkoski. 1982. Mass extinctions in the marine fossil record. *Science* 215:1501-03.

Sepkoski, J. J. 1982. A compendium of fossil marine families. *Milwaukee Public Museum Contributions in Biology and Geology*, no. 51, 125 pp.

Sepkoski, J. J., R. K. Bambach, D. M. Raup, and J. W. Valentine. 1981. Phanerozoic marine diversity and the fossil record. *Nature* 293:435-37.

Stanley, S. M. 1973. An explanation for Cope's Rule. *Evolution* 27:1-26.

Valentine, J. W. 1969. Patterns of taxonomic and ecological structure of the shelf benthos during Phanerozoic time. *Palaeontology* 12:684-709.

Van Valen, L. 1973. A new evolutionary law. *Evolutionary Theory* 1:1-30.

Yule, G. U. 1926. Why do we sometimes get nonsense-correlations between time-series? -- A study in sampling and the nature of time-series. *Journal of the Royal Statistical Society* 89:1-64.

On Replacing the Idea of Progress with an Operational Notion of Directionality

Stephen Jay Gould

Progress as Hindrance

Progress is a noxious, culturally embedded, untestable, nonoperational, intractable idea that must be replaced if we wish to understand the patterns of history. Yet our obsession with progress records something larger, deeper and vitally important in our search to understand the workings of time. Progress is a bad example of a crucial generality that we must pursue -- the study of directional change in history.

Progress slots into the logic of our cultural hopes as a response to scientific discoveries that we, as geologists and paleontologists, imposed upon an initially unwilling Western world. If life be stable on an earth but a few thousand years old, then human domination pervades history and we need no story of its gradual and inexorable development. But as soon as we learned that the earth is billions of years old, and that human history occupies but a metaphorical microsecond at the very end, then our central notion of human superiority in a meaningful world met its strongest challenge from science. Our geological confinement to a moment at the very end of recorded time must engender suspicions that we are a lucky accident, an afterthought rather than the goal of all creation. Progress is the doctrine that dispels this chilling thought -- for if life moves inexorably forward, however fitfully, towards its ultimate embodiment in human consciousness, then the restriction of *Homo sapiens* to a final moment poses no challenge to the general hope; for all that came before may now be interpreted as part of a process scheduled to yield our form from the start. We may then continue to pervade time through a long chain of imperfect surrogates, mounting their steady course towards our

GRANADA TV RENTAL'S
THEORY
OF EVOLUTION.

Figure l. The vernacular equation of the word evolution with progress. Reproduced with permission of Granada TV Rental.

exalted estate.

The continuing hold of progress upon our cultural perceptions may best be illustrated by a route too little exploited by scholars: the iconography of advertisements and cartoons, the leading visual styles of pop culture. I present just two examples from my large collection, anecdotes to be sure, but both well known to all of us as standard representations, not as idiosyncracies.

COMMENTARY

lany johnilil

THE EVOLUTION OF MAN

Figure 2. The evolution of man, by Larry Johnson, Boston Globe (before a Patriots-Raiders football game). The humor depends absolutely on the cartoonist's confidence that we will all immediately grasp the iconography of linear progress. Reproduced with permission of Larry Johnson.

The first (fig. 1) demonstrates that the very word "evolution" means progress in pop culture, for the ad makes no sense otherwise -- that is, one must believe that Granada's theory of evolution is progress in communication from rock to television, not simply (as professionals would define the term) a series of changes that respond to shifting local environments. The second (fig. 2) illustrates the confidence felt by cartoonists that everyone understands the image of a line of progress as a representation of evolution. Otherwise, no laughs.

In this paper, I wish to make three major points, all supporting the argument that we can preserve the deeper (and essential) theme of direction in history, while abandoning the intractable notion of progress: (1) progress is a culturally conditioned, not an inevitable or necessarily true account of history; (2) new themes in evolutionary theory have provided different interpretations of data conventionally treated as examples of progress; and (3) the larger issue of direction in history can be reformulated in a tractable manner.

The Power and Persistence of Nonprogressionist Visions

Far from being either an obvious truth of nature or a psychological necessity of all cultures, the idea of progress is a latecomer in one particular kind of society -- our own. Many historians from Bury (1920) to Lovejoy (1936) to Tawney (1926), have located the origin of progress (viewed as a general feature of both nature and human life) in the seventeenth century ferment of religious change, scientific discovery, and industrial innovation that signaled the spread and domination of Western culture throughout most of the world. Other scholars, including Eliade (1954) and Morris (1984), show that the opposite vision of an unchanging or strictly cycling earth is not only far older but has been, and continues to be, the general view of time maintained by most peoples.

As Eliade argues in his classic treatise, the notion of directional history motivated by human action is viewed as terrifying, not uplifting, by most cultures -- for such a concept of progress requires that direct human control over events be embraced, and we are then forced to admit that the pervasive tragedies of life, from plague to famine, are ordinary episodes under our influence and not (as the alternate view of stable immanence allows) anomalous moments in nature's constancy, and therefore subject to repeal or placation by prayer and sacrifice.

This restriction of progress may not impress readers, who might reply: since science is largely an invention of the culture that embraced progress, who cares about its unpopularity in other systems? But the limitation of progress goes further: many important Western scientific theories of time, life and history have also placed the denial of progress at the center of their conceptual structure. The "uniformitarian" geology of Hutton and Lyell provides the most striking of all examples. Our failure to grasp this philosophical basis underlies the standard misinterpretation that textbook cardboard presents for the supposed founders of our science (see my book *Time's Arrow, Time's Cycle,* 1987a, for an elaboration of this argument).

We view Hutton and Lyell as heroes of empirical science who transformed our view of the earth by inductive field work. In fact, the original and distinctive ideas, espoused by both and enshrined in our credo of uniformitarianism, arose prior to their field work and had a foundation in their

AWFUL CHANGES.

Man found only in a Fossil State.——Reappearance of Ichthyosauri.

A Lecture.—" You will at once perceive," continued Professor Ichthyosaurus, " that the skull before us belonged to some of the lower order of animals ; the teeth are very insignificant, the power of the jaws trifling, and altogether it seems wonderful how the creature could have procured food.'

Figure 3. De la Beche's mordant caricature of Lyell's belief in nondirectional cyclicity as the pattern of life's history. The returned ichthyosaur lectures on a product of the last creation.

idiosyncratic rejection of progress for a strict belief in the cyclic and nondirectional character of history. Hutton's cycles began so long ago that we see no vestige of beginning; they repeat with such precision that we can discern no prospect of an end so long as nature's current laws persist. Lyell anticipated the discovery of Paleozoic mammals to prove the unchanging mean complexity of life through time, and he even predicted that extinct genera would reappear when the appropriate stage of the grand climatic cycle, or great year, came round again. The famous caricature drawn by De la Beche (fig. 3) is not, as so long misinterpreted, a gentle satire on William Buckland's lectures, but a mordant dig from a classical progressionist at Lyell's view that ichthyosaurs would one day return (to lecture on a human skull of the "previous" creation -- see Rudwick 1975; Gould 1987a). For Lyell wrote (1830, p. 123):

Then might those genera of animals return, of which the memorials are
preserved in the ancient rocks of our continents. The huge iguanodon
might reappear in the woods, and the ichthyosaur in the sea, while the
pterodactyle might flit again through umbrageous groves of tree ferns.

Why We Need Different Explanations for the Phenomena of Progress

I do not deny that the fossil record contains legitimate cases of the
primary phenomenon identified as progress -- persistent trends within clades
based on characters interpretable as structural improvements, and leading to
increase in representation of taxa bearing these features (usually at the
expense of assumed competitors who don't). Every word in this definition
(excluding only articles and prepositions) can be challenged as ambiguous,
hence the extraordinary difficulty and contentiousness of the concept. (And I
have not even mentioned the primary issue of phylogenetics: shall we confine
the idea to trends within strictly monophyletic clades, or shall we also admit
parallel and iterative tendencies of several subclades within a monophyletic
group.)

Among the many caveats, consider just two primary items:

(1) A question of evidence: Our previous neo-Darwinian orthodoxies, now
dispersed, often led us to assert a claim of progress based on ambiguous data
that could yield no such firm interpretation without what my old logic teacher
used to call the "inarticulated major premises" (hidden assumptions to devotees
of the vernacular) of both strict adaptationism and the competitive basis of
faunal replacement. With these premises granted a priori, we falsely judged
two classes of data as evidence *ipso facto* for conventional ideas of progress.

First, we read the simple documentation of a general trend as evidence
for progress (what else, if it arose by anagenesis in a world dominated by
natural selection). With new ideas about the power of random processes and,
particularly, the pervasiveness of side consequences (the "spandrels" of Gould
and Lewontin 1979) in a revised world with expanded causes (species selection
with hitchhiking morphology, to cite just one example), we now understand
that simple documentation of a trend can only be the first step in asserting
progress as its basis. I suspect that many of the most famous trends of the
fossil record -- stipe reduction in graptolites (Elles 1922; Bulman 1963), or

increasing symmetry of the cup in crinoids (Moore and Laudon 1943), for example -- will be candidates for reinterpretation, especially since their assumed basis in morphological improvement has been elusive for so long (and may never have existed in our new perspective).

Second, we have taken faunal replacement (by forms judged sufficiently similar to share a general mode of life) as evidence *ipso facto* for improvement by competitive replacement. Clearly, our revised view of the greater frequency, speed, effect and causal distinctness of mass extinction must call this assumption into question, especially since so many of these classic replacements did not occur by gradual waning of one clade with coordinated waxing of its "successor," but by "relay" of one group to another at an extinction boundary (see Gould and Calloway 1980, on bivalves and brachiopods). Noncompetitive replacement may be the rule rather than the exception in life's history. I need hardly add that our own existence is probably contingent on such a replacement of dinosaurs by mammals.

(2) A question of definition: Do we consider a poker game progressive when players up the ante? Not usually, I think. The stakes are higher, but the rules don't change; a full house still beats a flush. I do not deny that coordinated change on this model does occur through time, especially when biological interactions are important (see Vermeij's elegant documentation of this theme -- 1973, 1987). Is a snail with a thick shell "better" than its thinner-shelled ancestor because an increase in the power of crushing predators requires this degree of strength to achieve the same adaptation that ancestors attained with thinner shells? The later world is different by virtue of such "arms races," but in what usual sense of the term can we proclaim it better?

As a historical footnote, William Buckland invoked this phenomenon to resolve an apparent conflict between progress and God's timeless perfection (see his 1836 Bridgewater Treatise on the power, wisdom, and goodness of God, as manifested in the Creation). Progress might seem to imply past imperfection, a concept inconsistent with the notion of a benevolent and omnipotent deity. But if each prior state was optimal in itself, and change only ups the ante without modifying the rules, then a continuous perfection can escalate in time.

Nonetheless, I do not deny that we have examples (probably many) of trends that meet our two major criteria for identification as progress in the usual vernacular sense: directional change fairly interpreted as biomechanical improvement (either in general design or in relation to a particular ecology continuously occupied by the group), and increasing relative representation in numbers, space or taxa as an inferred consequence of the trend. Lidgard (1986), for example, has traced the gradual increase in representation of the more efficient zooidal budding mode (at the expense of intrazooidal budding) through 100 million years in the evolution of encrusting cheilostome bryozoans. And Stanley's (1975) views on mantle fusion and subsequent formation of siphons as key innovations in the history of bivalves gains support from both criteria of biomechanical improvement and increasing relative representation.

Yet even these properly winnowed cases may undergo a radical reinterpretation in the light of new ideas stemming from the most important contemporary reformulation of evolutionary theory -- the hierarchical principle of interacting levels (Salthe 1985; Vrba and Eldredge 1984; Vrba and Gould 1986). The old notions of anagenesis and adaptation (both based on the assumption that trends are driven solely by natural selection operating on organisms) must be supplemented by new reasons for the phenomena of progress; these new reasons may be numerically predominant among the legitimate cases of progress.

I will present here only two revised notions of cause, one critique each for anagenesis and adaptation: (1) When we recognize that trends can be powered by the differential success of species, not only the struggles of organisms, we are led to view the lower level of selection within populations as part of an "entity-making machine," with trends formed at the higher level of sorting among entities. This suggests a different perspective on trends as changes in variance rather than anagenesis of entities. (2) Classical adaptation may be an impediment to, not the basis of, most evolutionary trends. The chief ingredient of a trend is the size and flexibility of the "exaptive pool," not the perfection of active and crucial adaptations.

Trends as changes in variance. Eldredge (1979) has argued that we must replace our traditional "transformational" concept of evolution with a "taxic" perspective. In this view, and adding the hierarchical theme, species (usually stable throughout their histories) are the entities that fashion trends by their differential births and deaths. A trend is the positive sorting of certain kinds of species, not the gradual transformation of a continuous population.

When we redefine trends as differential sorting of species, we may no longer interpret them as states (symmetry of a crinoid cup) with mean values gradually changing through time. The mean value of a clade is not a central adaptation, but only an abstraction, an average among many species each well adapted (by a particular state of the character in question) to its own environment. Means move because the distribution of variance changes. In ordinary directional selection within populations, we view an altered mean as an advantageous character and, therefore, as the direct instigator of changes in variance that secure its state through time (if larger mean size be favored by selection, then individuals so endowed leave more offspring on the average). But the story must be reversed for the taxic view: each species is an entity; they do not (for the most part) interact. The mean of a clade is an acausal abstraction, not the agent of anything. Means shift passively as a result of primary changes in variance produced by a differential sorting of species within the clade.

Trends, in other words, may be passive expressions of changes in variance -- misperceptions based on our conventional look at the wrong thing: abstracted averages rather than distributions. Suppose, for example, that we plot average distance from shore for marine species in a clade that originated in nearshore waters (not an abstract or randomly chosen *Gedanken* experiment, given the impressive literature asserting just this as an empirical regularity -- Jablonski et al. 1983, for example). We then plot the distribution of distances as the clade expands through time. Since new species are free to branch in only one direction (they are structurally precluded from invading land), the frequency distribution spreads out away from shore as time advances and the clade grows. The median value may never change -- we may always find half the species between shore and 50 meters out, and half in deeper water. But the mean will continually increase because the range of deeper species may expand, while the scope of the shallower half remains limited by the nearby shoreline.

I don't think that we would want to label this situation as a trend to increasing depth, yet we would so identify it by two standard and inappropriate measures: either by plotting an increase of the deepest species through time (and treating this abstraction as a continuous object, rather than an extreme variant of a system), or by calculating the mean value of the clade (and falsely thinking that we capture an unambiguous property of the whole). Instead, we witness an asymmetrical increase in variance, with a shifting mean as the skewness of the distribution expands, but no change in median. The asymmetry of increasing variation requires explanation of course, but of a sort quite different from usual suggestions of the anagenetic view. Our traditional perspective requires that we view the shifting mean itself as a causal basis of change. But here, the asymmetry of starting point is the cause we seek -- the curve was only free to expand in one direction. There never was an intrinsic advantage to deeper water (and no net movement in proportion of species in this direction), but only more room. The required explanatory principle is structural (a trivial point about geography of shore lines in this case), not adaptational. The trend is a change in variance and needs to be so interpreted.

I first developed this argument in my work on the history of batting averages in baseball (Gould 1983, 1986); I shall set forth its full context for our concerns in my presidential address to the Paleontological Society for 1987 ("Trends as changes in variance"). I shall, for now, only mention two examples to illustrate the extensive scope of this principle:

(a) Cope's rule, as Stanley suggested (1973), is a prime candidate for such explanation. There may be no general advantage in larger size, only more room in this direction *if* ancestral species tend to be small (as another of Cope's principles -- the so-called "law of the unspecialized" -- suggests). Since animals can't attain negative size, origin near the lower limits would provide little room for decrease, but great scope for increase. Median size may remain constant, as mean size grows. We cannot interpret such a circumstance as a meaningful trend to larger size in the usual sense, but only as movement into more space away from an asymmetrical starting point near one limit.

Jablonski (1987) has recently provided an example. He calculated sizes of both smallest and largest species within bivalve genera at their initial appearance and for their latest Cretaceous record. Of 58 genera, 33 did follow Cope's rule in the broad sense that the largest species of the latest Cretaceous exceeded the largest species at the point of origin. But only 11 of these genera showed a corresponding increase for the smallest species. For the

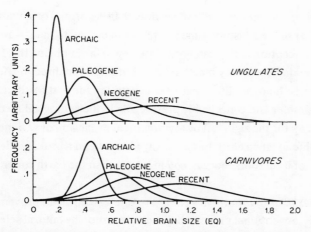

Figure 4. Frequency distributions for EQ (encephalization quotient) for carnivores and ungulates during the Tertiary. Note how the variance expands and how the region of small brains remains occupied throughout. From Jerison 1973.

others, size of the smallest species either remained stable or decreased. Jablonski concludes that "Cope's rule is driven by an increase in variance rather than a simple directional trend in body sizes."

(b) The undoubted trend to increasing brain size in many lineages of Tertiary mammals is often treated as a pure tale of selective advantage and anagenesis. But Jerison (1973), in the classic work on this subject, found (see fig. 4) that frequency distributions do not simply march up the ladder of brain size. Rather, ancestors had small brains, and the frequency distributions expand thereafter, while their left ends remain anchored at the small size of ancestral lineages. To be sure, the distributions do shift -- medians increase as well as means; but the major feature of change is a flattening and expansion in the range of the frequency distribution, not a general increase for all lineages within the clade. Jerison writes (1973, pp. 315-16):

> Diversity evolved just as average size evolved . . . . Despite the general trend toward increase in average brain size, there is an interesting and important overlap in the region of low brain size which indicates that there were at least some small-brained species present at all times. The evolution of enlarged brains, though generally a route to success and survival of new species, was not universal even among progressive orders.

I can't help adding, as a footnote to this section, that the primal observation of global progress in life's history -- the "march from monad to man" -- rests upon a biased perception in this category. Life began in simplicity; this fact of structure provided the "entity machine" with but one open direction, for complexity is a vast field, but little space exists between the first fossils, and anything both simpler and conceivably preservable in the geological record. Where else but up for the right tail by asymmetrical expansion of variance? But the modal life form on earth has probably always been a prokaryotic organism; each of us harbors more *E. coli* in his gut than the earth contains people.

Size and flexibility of the exaptive pool. We often attribute improvement of design to the conventional process of direct adaptation by natural selection; but this view cannot be correct in most cases. First of all, most complex adaptations are restrictive in their intricate match with environmental particulars; they correlate with low geological longevity in the light of changing climates and geographies. This, of course, is generally admitted. Instead, we usually seek the link of progress with adaptation in a subset of morphological novelties qualifying as "general adaptations" or biomechanical improvements in overall function. I believe that a logical error underlies this standard view.

If we understand adaptation aright, then even these general improvements arise for particular reasons in definite environments (legs for the possibility of land excursions in the search for more water, according to one classical view). They do not usually evolve "for" the general use that underlies their role in fostering the expansion of a large clade. The possibility of fostering is a question of structure, not immediate adaptation: how flexible is the new feature for the set of related uses required in a large clade; how appropriate is the rest of the *Bauplan* for adaptive expansion; how tight are the developmental ties between this new feature and other parts, and how dissociable for mosaic evolution --- to name just a few. The wings of insects could play such a fostering role; those of flying fishes could not. Yet I am not sure that the first winged insect was shaped any less than the first flying fish by narrow adaptation to a particular environmental role. We must focus our concern upon structural possibilities *following* the adaptation's original utility. Insects had evolved a new feature with flexibilities; flying fishes coopted a pectoral fin with other functions, and therefore limited in evolutionary malleability.

Exaptations (Gould and Vrba 1982) are features now useful to organisms (of adaptive value in the static and ahistorical sense of the term), but that originated for other reasons (therefore not by direct adaptation in the causal sense of the term favored from Darwin 1859, to Williams 1966). The broad domain of exaptation includes several categories, three most important for our altered understanding of the source of progress: (a) features evolved for one use but suited, by virtue of structure, for cooptation to another (this, and only this, phenomenon counts as "preadaptation" in our usual terminology; thus, preadaptation -- besides being a dreadful and confusing term that should be expunged -- is not a synonym for exaptation); (b) the expansion required by notions of hierarchy: features evolved for reasons, perhaps adaptive, at one level, but producing effects at the level of phenotypes; and (c) the expansion required by critiques of adaptation: features that never were adaptations, but arose as structural by-products of other features -- the spandrels of Gould and Lewontin, 1979.

The central attributes of evolving complexity, including human consciousness as a prominent example, must arise primarily from the last two sources of exaptation -- those furthest removed from any ordinary view of adaptation. Most evolutionists agree, for example, that gene duplication may be a prerequisite for evolutionary flexibility to build complexity (Ohno 1970), for some copies may continue to make required products while others become free to change. But duplicate genes did not evolve in any particular organism "for" a flexibility only advantageous to descendants -- unless our basic ideas of causality are awry. If they often evolve as "selfish DNA" by genic selection (an attractive modern theory, see Orgel and Crick 1980; Doolittle and Sapienza 1980), then they are exaptations for organisms by theme (b). If we assume, as I believe we must (Gould and Vrba 1982; Gould 1987b), that most basic features of human consciousness arise as spandrels in the construction of such computing power, then most distinctively human traits are exaptations by theme (c).

One of the most powerful intellectual ideas of our century, Chomsky's theory of language, is an explicit argument about spandrels though varying terminologies and lack of communication across disciplines have obscured this fact (Chomsky 1986 and personal conversations). Chomsky holds that the deep structure of language is universal and innate as an organ of the brain. It arises in normal development without special nurturing, so long as environments be adequate, just as any organ of morphology does. It works well, but it is too quirky and nonoptimal, too much a historical particular rather than a predictable mechanism, to be viewed as something gradually and directly evolved for function. Chomsky prefers to view deep structure as something coopted *in toto* from other sources -- as a spandrel, originally evolved for other cognitive roles (chimps, lacking language entirely so far as we know, are richly endowed with cognition and intelligence by any meaning of the term). Chomsky's argument often strikes evolutionists as absurd, if not heretical; we are too long trained to view language as the immediate source of human triumph. But, given Chomsky's remarkable track record in prediction and efficacy (his insistence, for example, now confirmed, that signing experiments on chimps had been misinterpreted and were not teaching language), I believe that evolutionists must weigh his views seriously.

Thus, exaptive possibilities, not adaptive usages, form the ground of progress, the source of flexibility in change (including the potential to build upon existing structure towards a new state that we judge as more complex). The possibility of progress is specified by the size and flexibility of the "exaptive pool" -- the set of cooptable features. Progress is not regulated by adaptation, especially in the light of new views that locate so much of the exaptive pool, not in features that first arose as adaptations of phenotypes (classical preadaptations), but either at other levels of the hierarchy, or as nonadaptations at the phenotypic level. The rules of structure and development, not the workings of adaptation, set the boundaries and possibilities of progress.

The Deeper Issue, Restrictively and Misleadingly Addressed by Progress, Can be Reformulated in a Tractable Way

We have already seen how Hutton's rigidly cyclical view of time precluded progress entirely. But it did more, and to regrettable effect. It also ruled out the larger subject that progress represents as a poor and limited example

-- history itself. Hutton's geology is a curiously restricted subject to modern readers, for Hutton, in trying to make the earth cycle through time as it moves through Newton's space, denied history in any meaningful sense of the term (Gould 1987a). What then does history require?

Two central attributes mark an interpretable history. First, events must have temporal signatures; they must represent their moment distinctly. Otherwise, history is not a sequence of definable events, since any episode may recur precisely and we cannot, therefore, know where we are. (Hutton's geology denied history in this most fundamental sense, since his precise cyclicity granted no uniqueness to any moment.)

But distinctness is not enough, at least in our culturally conditioned view of intelligibility. History must be more than a string of isolated, if distinctive, events strung together one after the other. We view history as its last five letters -- a story, a skein unwinding in some particular, if complicated, direction. A temporal sequence without directionality, and without causal links among its events, is not inconceivable (the world could be so fashioned), but neither would we call such a thing history. History needs both distinctiveness and directionality to meet our definitions and pique our interest.

What the science of history requires above all is a tractable way to study directional processes. This, I believe, is the proper reason why we have been so obsessed with the idea of progress (I leave aside the improper reason of our cultural hopes and traditions). The concept of directionality, or temporal vectors, is the core of our legitimate concern, for historical science requires its validation. We have been led astray by sinking a proper concern with directionality entirely into one possible and limited manifestation as progress. Thus, we have compromised the deeper theme by relying upon a poor and biased example to carry the richness of the entire subject. What we need is a methodology for studying time's directions.

I have recently made this argument *in extenso* (Gould et al. 1987), and will only present a sketch and an example here. Physical scientists have confronted the paradox that so little in nature's laws imparts a directionality to time, despite our overwhelming impression that such vectors exist. Beyond the second law of thermodynamics, "time's arrow" in Eddington's phrase, we can point to little else without ambiguity or subjectivity (see Morris 1984). Morris suggests a base-level definition for temporal arrows: "we mean only

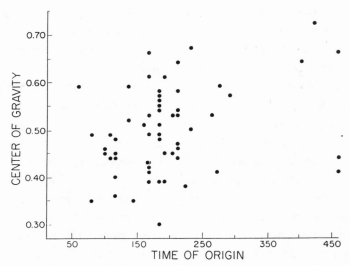

Figure 5. Families within higher taxa. Center of gravity plotted as a function of time of origin from the base of the Vendian for extinct groups. General rise in value shows that bottom-heavy clades are concentrated among those that originate early. From Gould et al. 1987.

that the world has a different appearance in one direction of time than it does in the other." In other words, we must look for temporal asymmetry. Expressed in our terms and by modern metaphor: if I hand you the tape of life's history but do not tell you which end is which, would you know, in watching the tape unfold, whether it was running properly forward or illegitimately backward? This may seem a risibly small question compared with (genera within families) throughout the grandiose idea of progress, but it is definable, tractable, testable, and quantifiable. In other words, we may treat this approach to directionality as science rather than as an ambiguous instantiation of our hopes -- a good trade.

In our previous work (Gould et al. 1987), we looked for temporal asymmetry in the standard paleontological representation of life's temporal structure: clade diversity diagrams. We asked the simple operational question: if I hand you a chart of clade diversity diagrams but forget to label the vertical axis of time, would you know which end was Cambrian, which Recent? We found, by several criteria, that a temporal arrow could be defined by the

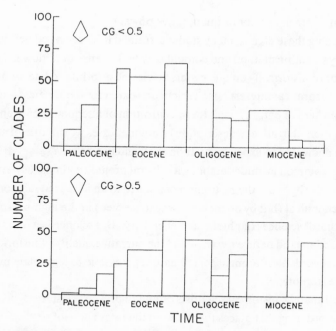

Figure 6. Histograms for bottom-heavy (above) and top-heavy (below) clades of extinct Tertiary mammals. Note that early times featured a higher proportion of bottom-heavy clades. CG = Center of Gravity. From Gould et al. 1987.

bottom-heavy character of clades that arose early in the history of a larger group (bottom-heavy clades concentrate their times of greatest flourishing before the midpoint of their geological range; clades arising later show no directionality, and maintain centers of gravity at their midpoints, on average). We affirmed this pattern both for the history of marine invertebrates (genera and families) and for the Tertiary history of mammalian genera (figs. 5 and 6). We also argued that its apparently fractal character indicated a possible generality, or true arrow of evolutionary time. (The bottom-heavy character of much larger monophyletic clades indicates a preservation across levels -- consider the 20 or so Paleozoic echinoderm classes, versus five today; or the history of the coelomate Metazoa, by comparing the astounding diversity of the Burgess Shale with current stereotypy.) We also argued that the method of clade diversity diagrams could be extended to any field that studies changing amounts and percentages through time (it has long been used in archaeology, but without appreciation for its quantitative power).

In reading these claims, many students trained in stereotyped definitions of a unitary scientific method and content may feel a sense of letdown. How can a vector of historically unique events be viewed as primary data, or as bearers of theoretical interest? Isn't science a search for the timeless and quantifiable laws of nature? Isn't the specification of a sequence of events no more than mere narrative, a description of uniquenesses worth little in an enterprise dedicated to experiment, repetition and prediction?

These common attitudes embody a cultural prejudice almost as pervasive as the idea of progress -- the restriction of science to an idealized method touted as canonical (but by no means always observed) in the high-prestige, so-called "hard" sciences of physics and chemistry. But science is a pluralistic search to understand nature's ways -- and the narrative quality of historical sequences records a different aspect of nature accessible to legitimate methods beyond the stereotype.

Unfortunately, historical scientists often fall prey to the stereotype, and to the rank-ordering that places them low on the totem pole of reputation. We have often made the mistake of trying to ape the inappropriate methods, or bowing to the supposed data, of our colleagues in fields with greater prestige. Thus, Hutton denied history and tried to formulate a geology that would emulate Newton's timeless cosmos. Kelvin proclaimed an impossibly young earth, and most geologists accepted his authority, even though the data of sequential history pointed to times much longer. Charles Spearman invented factor analysis in order to misidentify intelligence as a single entity within the brain, and then touted his quantity as the key to a scientific psychology: "This Cinderella among the sciences has made a bold bid for the level of triumphant physics itself" (see Gould 1981).

The sequential events of historical narrative provide the primary data for our paleontological science of life's history, and our evolutionary science of phylogeny. We can accord these data their due respect, while remaining true to the basic definition of science as a search for natural order, by insisting that the neglected subject of directionality in time become a focal point in the study of history.

References

Buckland, W. 1836. *Geology and mineralogy considered with reference to natural theology*. London: G. Routledge & Co.

Bulman, O. M. B. 1963. The evolution and classification of the Graptoloidea. *Quart. Jour. Geol. Soc. London* 119:401-18.

Bury, J. B. 1920. *The idea of progress*. London: Macmillan.

Chomsky, N. 1986. *Knowledge of language: Its nature, origin and use*. New York: Praeger.

Darwin, C. 1859. *On the origin of species*. London: John Murray.

Doolittle, W. F., and C. Sapienza. 1980. Selfish genes, the phenotype paradigm and genome evolution. *Nature* 284:601-3.

Eldredge, N. 1979. Alternative approaches to evolutionary theory. In *Models and methodologies in evolutionary theory*, ed. J. H. Schwartz and H. B. Rollins. *Bull. Carnegie Mus. Nat. Hist.* 13:7-19.

Eliade, M. 1954. *The myth of the eternal return*. Princeton: Princeton University Press.

Elles, G. L. 1922. The graptolite faunas of the British Isles. A study in evolution. *Proc. Geologists' Assoc.* 33:168-200.

Gould, S. J. 1981. *The mismeasure of man*. New York: W. W. Norton.

Gould, S. J. 1983. Losing the edge: The extinction of the .400 hitter. *Vanity Fair* (March):120, 264-78.

Gould, S. J. 1986. Entropic homogeneity isn't why no one hits .400 anymore. *Discover* (August): 60-66.

Gould, S. J. 1987a. *Time's arrow, time's cycle*. Cambridge: Harvard University Press.

Gould, S. J. 1987b. *An urchin in the storm*. New York: W. W. Norton.

Gould, S. J., and C. B. Calloway. 1980. Clams and brachiopods -- ships that pass in the night. *Paleobiology* 6:383-96.

Gould, S. J., N. L. Gilinsky, and R. Z. German. 1987. Asymmetry of lineages and the direction of evolutionary time. *Science* 236:1437-41.

Gould, S. J., and R. C. Lewontin. 1979. The spandrels of San Marco and the Panglossian paradigm: A critique of the adaptationist programme. *Proc. R. Soc. London* B 205:581-98.

Gould, S. J., and E. S. Vrba. 1982. Exaptation -- A missing term in the science of form. *Paleobiology* 8:4-15.

Jablonski, D. 1987. How pervasive is Cope's Rule? A test using Late Cretaceous mollusks. *Geol. Soc. America. Abstracts with Programs* 19:713-14.

Jablonski, D., J. J. Sepkoski, Jr., D. J. Bottjer and P. M. Sheehan. 1983. Onshore-offshore patterns in the evolution of Phanerozoic shelf communities. *Science* 222: 1123-25.

Jerison, H. J. 1973. *Evolution of the brain and intelligence*. New York: Academic Press.

Lidgard, S. 1986. Ontogeny in animal colonies: A persistent trend in the bryozoan fossil record. *Science* 232:230-32.

Lovejoy, A. O. 1936. *The great chain of being*. Cambridge: Harvard University Press.

Lyell, C. 1830. *The principles of geology, Vol. 1*. London: John Murray.

Moore, R. C ., and L. R. Laudon. 1943. Evolution and classification of Paleozoic crinoids. *Geol. Soc. America Spec. Pap.* 46, 153 pp.

Morris, R. 1984. *Time's arrows*. New York: Simon and Schuster.

Ohno, S. 1970. *Evolution by gene duplication*. New York: Springer.

Orgel, L. E., and F. H. C. Crick. 1980. Selfish DNA: The ultimate parasite. *Nature* 284:604-07.

Rudwick, M. J. S. 1975. Caricature as a source for the history of science: De la Beche's anti-Lyellian sketches of 1831. *Isis* 66:534-60.

Salthe, S. V. 1985. *Evolving hierarchical systems*. New York: Columbia University Press.

Stanley, S. M. 1973. An explanation for Cope's Rule. *Evolution* 27:1-26.

Stanley, S. M. 1975. Adaptive themes in the evolution of the Bivalvia (Mollusca). *Ann. Rev. Earth Planetary Sci.* 3:361-85.

Tawney, R. H. 1926. *Religion and the rise of capitalism*. New York: Harcourt, Brace and Co.

Vermeij, G. J. 1973. Adaptation, versatility, and evolution. *Syst. Zool.* 22:466-77.

Vermeij, G. J. 1987. *Evolution and escalation. An ecological history of life*. Princeton: Princeton University Press.

Vrba, E. S., and N. Eldredge. 1984. Individuals, hierarchies and processes: Towards a more complete evolutionary theory. *Paleobiology* 10:146-71.

Vrba, E. S., and S. J. Gould. 1986. The hierarchical expansion of sorting and selection: Sorting and selection cannot be equated. *Paleobiology* 12:217-28.

Williams, G. C. 1966. *Adaptation and natural selection*. Princeton: Princeton University Press.

Contributors

Francisco José Ayala is Distinguished Professor in the Department of Ecology and Evolutionary Biology at the University of California, Irvine, CA 92717. A recognized authority on the philosophy of biology, he is a member of the National Academy of Sciences and the American Academy of Arts and Sciences and past president of the Society for the Study of Evolution. He is best known for his work on genetics and is an author or editor of ten books and over 300 scientific papers.

Robert C. Dunnell, Professor of Anthropology in the Department of Anthropology, University of Washington, Seattle, WA 98195, is also Adjunct Professor of Quaternary Studies and Adjunct Curator of North American Archaeology at the Washington State Museum as well as Affiliate Curator, Department of Anthropology, Peabody Museum of Natural History, Yale University. He is a fellow of the American Association for the Advancement of Science, and the Society of American Archaeology representative of its Section H Executive Committee. He is also president of the Association for Field Archaeology. He is the author or editor of three books and numerous articles, principally in the area of archaeological classification, dating methods, and the prehistory of the United States.

Stephen J. Gould, Alexander Agassiz Professor of Zoology and Professor of Geology at the Museum of Comparative Zoology, Harvard University, Cambridge, MA 02138, is also a member of the Committee of Professors, the Department of Organismic and Evolutionary Biology, and the Department of History of Science. Gould is a recipient of numerous academic medals and awards, and a member of the American Academy of Arts and Sciences. He is past president of the American Society of Naturalists and president of the

Paleontological Society. In addition Gould has several honorary doctorates and literary awards. He has published ten books and more than 200 articles on biometrics, paleontology, evolutionary biology and the history of science. His particular interest is in the role of ontogeny in evolution, the nature of adaptation and in the causes of patterns of cladogenesis.

David L. Hull, Professor in the Department of Philosophy, Northwestern University, Evanston, IL 60208, is also acting chairman of the Department of Ecology and Evolutionary Biology at Northwestern. Hull is past president of the Society of Systematic Zoology and the Philosophy of Science Association, and an editor of the prestigious Science and its Conceptual Foundations Series of the University of Chicago. He is an author of three books and over 100 papers on philosophy and sociology of science, particularly evolutionary biology and systematics.

Thomas C. Kane is Associate Professor in the Department of Biological Sciences, University of Cincinnati, Cincinnati, OH 45221. He has also been Visiting Associate Professor in the Department of Ecology and Evolutionary Biology, Northwestern University, Evanston, IL, and, through grant support from the Centre National de la Recherche Scientifique, he has served as Visiting Associate Director of Research at the Laboratoire Souterrain, Moulis, France. His research on the ecology and evolution of cave animals has been funded by the American Philosophical Society, the Cave Research Foundation, Sigma Xi, and the National Science Foundation. He has published over twenty professional papers on this work.

John Maynard Smith, Professor Emeritus and past dean of Biology in the Department of Population and Plant Biology at the University of Sussex, Brighton, BN1 9Q6, U.K., was trained as an aeronautical engineer, and worked as an engineer during World War II. He began his biological work in functional morphology and is best known for his studies of evolutionary biology, particularly evolutionary genetics. Maynard Smith is a member of the

Royal Society and a foreign associate of the National Academy of Sciences, a recipient of the Darwin Medal of the Royal Society, and has honorary doctorates from Kent, Oxford, and Chicago universities. He is president-elect of the Society for the Study of Evolution and the author or editor of six books and over 250 scientific papers.

Matthew H. Nitecki is Curator of Fossil Invertebrates in the Department of Geology, Field Museum of Natural History, Chicago, IL 60605, and in the Committee on Evolutionary Biology at the University of Chicago. He also holds a faculty appointment in the College of the University of Chicago. Nitecki has written or edited about 150 papers, including ten books, on paleobiology of Paleozoic fossils, history and sociology of science, and theoretical evolutionary biology.

William B. Provine, Professor of the History of Biology in the Division of Biological Sciences, Section of Ecology and Systematics, Cornell University, Ithaca, NY 14853, also holds an appointment in the Department of History. His research on history of genetics, particularly of the twentieth century, and on the philosophical aspects of modern biology, and mechanics of evolution and grand designs in nature, has culminated in three books and numerous papers. He is also author of *Sewall Wright and Evolutionary Biology.*

David M. Raup is Sewell L. Avery Distinguished Service Professor in the Department of the Geophysical Sciences, University of Chicago, 5734 S. Ellis Ave., Chicago, IL 60637, and professor in the Committees on Evolutionary Biology and the Conceptual Foundations of Science. He also holds a research appointment at the Field Museum. He is past president of the Paleontological Society and a member of the National Academy of Sciences. Raup has for many years pioneered the application of mathematical approaches to the study of fossils and the history of life on earth, and has done original theoretical work on the morphology, growth, and phylogeny of living and fossil organisms. He has written or edited five books and over 150 papers.

Robert J. Richards is Associate Professor in the departments of History, Philosophy, and Biological Psychology at the University of Chicago, 1126 East 59th St., Chicago, IL 60637. He is chairman of the Committee on the Conceptual Foundations of Science, a history and philosophy of science doctoral program at Chicago. Richards has written extensively on the history and philosophy of biology and psychology, and has lectured on these subjects throughout the United States, in Europe, and in South America. He is the author of the recently published book *Darwin and the Emergence óf Evolutionary Theories of Mind and Behavior.*

Robert C. Richardson is currently Head and Professor of Philosophy in the Department of Philosophy at the University of Cincinnati, Cincinnati, OH 45221. Since receiving his Ph.D. in 1976 from the University of Chicago, Richardson has published broadly in philosophy of science, philosophy of biology, philosophy of psychology, and philosophy of mind. His collaboration with Thomas C. Kane began in 1982, and they are currently engaged in a joint project studying the rise and fall of American Neo-Lamarckism.

Michael Ruse, Professor of History and Philosophy in the Department of Philosophy at the University of Guelph, Guelph, ON N1G 2W1, Canada, is an authority on the philosophy and history of evolutionary biology, particularly Darwin. He is the author of numerous papers and several books, including a trilogy on Darwinism: *The Darwinian Revolution* (University of Chicago Press 1979); *Darwinism Defended* (Addison-Wesley 1982); and *Taking Darwin Seriously* (Blackwell 1986). Ruse is a fellow of the Royal Society of Canada.

Jeffrey C. Schank is a graduate student in the Committee on the Conceptual Foundations of Science at the University of Chicago, 1126 East 59th St., Chicago, IL 60637. He is primarily interested in problems arising in the analysis of model building in the biological sciences.

Adam Urbanek is Vice-President of the Polish Academy of Sciences, P. O. Box 24, 00-901 Warszawa, Poland, and a past Chairman of the Academy's Committee on Evolutionary and Theoretical Biology. He also teaches evolutionary biology at the University of Warsaw. He has initiated novel studies of fossil graptolites. During one of his visits to America he was a Visiting Scientist at the Field Museum. His publications on all aspects of paleobiology, evolutionary biology and philosophy of science include many monographs and over fifty papers.

E. O. Wiley is a Professor of Systematics and Ecology, and a Curator of the Museum of Natural History, University of Kansas, Lawrence, KS 66045. He is also a Research Associate, U.S. National Museum of Natural History, and a past president of the Society of Systematic Zoology. Wiley has authored two books (one coauthored with D. R. Brooks), and coedited (with C. Hocutt) one book. He has published numerous papers on the theory and practice of phylogenetic systematics, theoretical evolutionary biology, and the evolution and systematics of fishes.

William C. Wimsatt is Professor in the Department of Philosophy, University of Chicago, 1050 East 59th St., Chicago, IL 60637. He is also a professor in the Committees on the Conceptual Foundations of Science, Evolutionary Biology, General Studies in Humanities and the Fishbein Center for the History of Science and Medicine, and was a University Visiting Distinguished Professor at Ohio State University. Wimsatt has given lectures and seminars in the United States, Canada, Mexico, England and Belgium, and has published over twenty papers on a cluster of problems arising in the analysis of the structure, behavior, and evolution of complex functionally organized systems, and on the strategies and biases of mathematical modelling in evolutionary biology.

Index